# WORKSHEETS
## WITH THE MATH COACH

### GEX, INCORPORATED

# PREALGEBRA
## FIFTH EDITION

## Jamie Blair
*Orange Coast College*

## John Tobey
*North Shore Community College*

## Jeffrey Slater
*North Shore Community College*

## Jennifer Crawford
*Normandale Community College*

**PEARSON**

Boston   Columbus   Indianapolis   New York   San Francisco   Upper Saddle River
Amsterdam   Cape Town   Dubai   London   Madrid   Milan   Munich   Paris   Montreal   Toronto
Delhi   Mexico City   Sao Paulo   Sydney   Hong Kong   Seoul   Singapore   Taipei   Tokyo

Copyright © 2012, 2010, 2006 Pearson Education, Inc.
Publishing as Pearson, 75 Arlington Street, Boston, MA  02116.

ISBN-13: 978-0-321-78060-7
ISBN-10: 0-321-78060-4

1 2 3 4 5 6 BRR 15 14 13 12 11


www.pearsonhighered.com

PEARSON

**Worksheets with the Math Coach**

*Prealgebra, Fifth Edition*

# Table of Contents

**Chapter 1 Whole Numbers and Introduction to Algebra**
**1.1 Understanding Whole Numbers**

**Vocabulary**
whole numbers • digits • place-value system • period • round-off place • place-holder
expanded notation • word names • number line • inequality symbol • rounding

1. The whole numbers 0, 1, 2, 3, 4, 5, 6, 7, 8, 9 are called _____.

2. When we write very large numbers, we place a comma after every group of three digits, called a(n) _____, moving from right to left.

3. A number system in which the position, or placement, of the digits in a number tells the value of the digits is called a(n) _____.

4. _____ is a process that approximates a number to a specific round-off place.

| **Example** | **Student Practice** |
|---|---|
| **1.** In the number 573,025: | **2.** In the number 4,905,817: |
| **(a)** In what place is the digit 7? | **(a)** In what place is the digit 9? |
| 5<u>7</u>3,025 | |
| The digit 7 is in the ten thousands place. | |
| **(b)** In what place is the digit 0? | **(b)** In what place is the digit 5? |
| The digit 0 is in the hundreds place. | |
| **3.** Write 1,340,765 in expanded notation. | **4.** Write 5,302,769 in expanded notation. |
| Write 1 followed by a zero for each of the remaining digits. We continue in this manner for each digit. Since there is a zero in the thousands place, it is not included in the sum. | |
| $\underline{1},340,765$ as $\underline{1},000,000 + \underline{3}00,000 +$ $\underline{4}0,000 + \underline{7}00 + \underline{6}0 + \underline{5}$ | |

Vocabulary Answers: 1. digits  2. period  3. place-value system  4. rounding

| Example | Student Practice |
|---|---|
| **5.** Jon withdraws $493 from his account. He requests the minimum number of bills in one-, ten-, and hundred-dollar bills. Describe the quantity of each denomination of bills the teller must give Jon. | **6.** Michael withdraws $389 from his account. He requests the minimum number of bills in one-, ten-, and hundred-dollar bills. Describe the quantity of each denomination of bills the teller must give Michael. |

**5.** (continued)

If we write $493 in expanded notation, we can easily describe the denominations needed.

$$400 \quad + \quad 90 \quad + \quad 3$$
$$4 \qquad\quad 9 \qquad\quad 3$$

Thus, the teller must give Jon 4 hundred-dollar bills, 9 ten-dollar bills, and 3 one-dollar bills.

| Example | Student Practice |
|---|---|
| **7.** Write a word name for each number. | **8.** Write a word name for each number. |
| **(a)** 2135 | **(a)** 8592 |

**7. (a)** (continued)

Look at a place value chart if you need help identifying the period.

The number begins with 2 in the thousands place.

The word name is two thousand, one hundred thirty-five. Notice that we use a hyphen in "thirty-five."

**(b)** 300,460

**8. (b)** 5,230,089

**7. (b)** (continued)

The number begins with 3 in the hundred thousands place.

The word name is three hundred thousand, four hundred sixty.

We place a comma here to match the comma in the number.

| Example | Student Practice |
|---|---|
| **9.** Replace each question mark with the inequality symbol < or >.<br><br>(a) 1 ? 6<br><br>1 is less than 6.<br><br>1 < 6<br><br>(b) 8 ? 7<br><br>8 > 7 | **10.** Replace each question mark with the inequality symbol < or >.<br><br>(a) 3 ? 7<br><br><br><br>(b) 9 ? 5 |
| **11.** Rewrite using numbers and an inequality symbol.<br><br>(a) Five is less than eight.<br><br>Five  is less than  eight.<br> ↓      ↓      ↓<br> 5     <     8<br><br>(b) Nine is greater than four.<br><br>9 > 4 | **12.** Rewrite using numbers and an inequality symbol.<br><br>(a) Four is less than nine.<br><br><br><br>(b) Eight is greater than two. |
| **13.** Round 57,441 to the nearest thousand.<br><br>The round-off place digit is in the thousands place.<br><br>Identify the round-off place digit, 5⃞7⃞,441.<br><br>The digit to the right of 7 is less than 5, 5⃞7⃞,4̲41.<br><br>Do not change the round-off place digit. Replace all digits to the right with zero, 57,000. | **14.** Round 65,745 to the nearest hundred. |

3

| Example | Student Practice |
|---|---|
| **15.** Round 4,254,423 to the nearest hundred thousand. | **16.** Round 5,678,231 to the nearest ten thousand. |

The round-off place digit is in the hundred thousands place.

Identify the round-off place digit, $4,\boxed{2}54,423$.

The digit to the right of 2 is 5 or more, $4,\boxed{2}54,423$.

Increase the round-off place digit by 1. Replace all digits to the right with zeros.

$4,300,000$

Thus, 4,254,423 rounded to the nearest thousand is 4,300,000.

**Extra Practice**

**1.** Write the word name for $80,059$. Then write the number in expanded notation.

**2.** Replace the question mark with < or >.

$11,032 ? 10,032$

**3.** Rewrite using numbers and inequality symbols.

Ninety-eight is less than one hundred fourteen

**4.** Round 2940 to the nearest hundred.

**Concept Check**

Explain how to round 8937 to the nearest hundred.

Name: _____     Date: _____

Instructor: _____     Section: _____

**Chapter 1 Whole Numbers**
**1.2 Adding Whole Number Expressions**

**Vocabulary**
addition • addends • sum • variable • algebraic expression • variable expression
identity property of zero • addition facts • commutative property of addition • square
associative property of addition • evaluate • rectangle • perpendicular • triangle
right angle • simplify • perimeter • inductive reasoning

1.  A letter that represents a number is called a(n) _____.

2.  The _____ states that two numbers can be added in either order to produce the
    same result.

3.  To _____ an algebraic expression, we replace the variables in the expression
    with their corresponding values and simplify.

4.  The distance around an object is called the _____.

| **Example** | **Student Practice** |
|---|---|
| **1.** Translate the English phrase using numbers and symbols. | **2.** Translate the English phrase using numbers and symbols. |
| **(a)** The sum of six and eight | **(a)** Six increased by two |
| The words "the sum of" indicate addition, " + ." | |
| $6 + 8$ | |
| **(b)** A number increased by four | **(b)** The sum of a number and ten |
| A number  increased by  four $\downarrow$  $\downarrow$  $\downarrow$ $x$  $+$  $4$ | |
| Although we used the variable $x$ to represent the unknown quantity, any letter could have been used. | |

Vocabulary Answers: 1. variable  2. commutative property of addition  3. evaluate  4. perimeter

| Example | Student Practice |
|---|---|
| **3.** Express 4 as a sum of two whole numbers. Write all possibilities. How many addition facts must we memorize? Why? | **4.** Express 6 as a sum of two whole numbers. Write all possibilities. How many addition facts must we memorize? Why? |

**3.** (continued)

Write all the sums equal to 4 and observe any patterns.

$$4+0=4$$
$$3+1=4$$
$$2+2=4$$
$$1+3=4$$
$$0+4=4$$

Notice that the last two rows of the pattern are combinations of the same numbers listed in the first two rows. We need to learn only 2 addition facts for the number 4, $3+1$ and $2+2$. The remaining facts are either a repeat of these or use the addition property of zero.

| Example | Student Practice |
|---|---|
| **5.** Use the associative property and/or commutative property as necessary to simplify the expression $5+(n+7)$. | **6.** Use the associative property and/or commutative property as necessary to simplify the expression $(8+n)+4$. |

**5.** (continued)

Apply the commutative property.

$$5+(n+7)=5+(7+n)$$

Regroup using the associative property and simplify. Then rewrite with the variable first.

$$5+(7+n)=(5+7)+n$$
$$=12+n$$
$$=n+12$$

| Example | Student Practice |
|---|---|
| **7.** Evaluate $x + y + 3$ for the given values of $x$ and $y$. | **8.** Evaluate $x + y + 5$ for the given values of $x$ and $y$. |
| $x$ is equal to 6 and $y$ is equal to 1 | $x$ is equal to 9 and $y$ is equal to 14 |
| $x + y + 3$ $6 + 1 + 3$ $\quad 10$ | |

**9.** A market research company surveyed 1870 people to determine the type of beverage they order most often at a restaurant. The results of the survey are shown in the table. Find the total number of people whose responses were iced tea, soda, or coffee.

| Type of Beverage | Number of Responses |
|---|---|
| Soda | 577 |
| Orange juice | 475 |
| Coffee | 84 |
| Iced tea | 357 |
| Milk | 286 |
| Other | 91 |

We add whenever we must find the "total" amount. First add in the ones column, $7 + 7 + 4 = 18$. Since 18 is 1 ten and 8 ones, carry the tens by placing a 1 at the top of the tens column. Repeat the process for the tens and hundreds columns to complete the addition.

$$
\begin{array}{r}
{}^{2\ 1}\\
357\\
577\\
+\ \ 84\\
\hline
1018
\end{array}
$$

A total of 1018 people responded iced tea, soda, or coffee.

**10.** A market research company surveyed 1738 people to determine the type of meal they order most often at a restaurant. The results of the survey are shown in the table. Find the total number of people whose responses were fish, beef, or chicken.

| Type of Meal | Number of Responses |
|---|---|
| Beef | 678 |
| Chicken | 598 |
| Fish | 271 |
| Vegetarian | 191 |

| **Example** | **Student Practice** |
|---|---|

**11.** Find the perimeter of the triangle. (The abbreviation "ft" means feet.)

We add the lengths of the sides to find the perimeter.

$$5 \text{ ft} + 5 \text{ ft} + 7 \text{ ft} = 17 \text{ ft}$$

The perimeter is 17 feet.

**12.** Find the perimeter of the triangle. (The abbreviation "m" means meters.)

**Extra Practice**

**1.** Use the indicated property to rewrite the sum, then simplify if possible.

$(a+2)+7$ ; associative property of addition

**2.** Evaluate the expression using the given values of $x$ and $y$.

$x + 200 + y$ when $x = 645$ and $y = 1000$

**3.** Add.

3754
358
+6839

**4.** Find the perimeter of the figure.

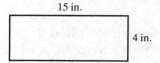

**Concept Check**

(a) When we carry, what is the value of the 1 that is placed above the 9?

(b) When we carry, what is the value of the 1 that is placed above the 3?

$$\begin{array}{r} {\scriptstyle 1\ 1} \\ 395 \\ +\ \ 28 \\ \hline 423 \end{array}$$

Name: _____     Date: _____
Instructor: _____     Section: _____

**Chapter 1 Whole Numbers**
**1.3 Subtracting Whole Number Expressions**

**Vocabulary**
subtraction • minus sign • minuend • subtrahend • difference • borrowing

1. The number being subtracted in a subtraction problem is called the _____.

2. The result of a subtraction problem is called a _____.

3. The symbol used to indicate subtraction is called a _____ "−."

4. We do _____ when we take objects away from a group.

| **Example** | **Student Practice** |
|---|---|
| **1.** Subtract. | **2.** Subtract. |
| (a) $9-5$ | (a) $8-5$ |
| $\quad 9-5=4$ | |
| (b) $15-0$ | (b) $19-19$ |
| $\quad 15-0=15$ | |
| **3.** $800-50=750.$ Use this fact to find $800-53.$ | **4.** $700-50=650.$ Use this fact to find $700-53.$ |
| Since we know $800-50=750,$ we can use subtraction patterns to find $800-53.$ | |
| $800-50=750$ <br> $800-51=749$ <br> $800-52=748$ <br> $800-53=747$ | |
| Observe the patterns in the second and the last columns. In the second column, the numbers increase by 1 and in the third column, they decrease by 1. | |

Vocabulary Answers: 1. subtrahend 2. difference 3. minus sign 4. subtraction

| Example | Student Practice |
|---|---|
| **5.** Translate using numbers and symbols.<br><br>Four less than seven<br><br>The words "less than" indicate subtraction, "−". Note that the numbers are reversed because of the way the phrase is worded.<br><br>$7 - 4$ | **6.** Translate using numbers and symbols.<br><br>The difference of eight and five |
| **7.** Evaluate $7 - x$ for $x = 2$.<br><br>Replace $x$ with 2 and simplify.<br><br>$7 - x = 7 - 2 = 5$<br><br>When $x$ is equal to 2, $7 - x$ is equal to 5. | **8.** Evaluate $9 - a$ for $a = 5$. |
| **9.** Subtract $304 - 146$.<br><br>We must borrow since we cannot subtract 6 ones from 4 ones.<br><br>Also we cannot borrow a ten since there are 0 tens, so we must borrow from 3 hundreds.<br><br>Rewrite 3 hundreds as hundreds and tens, $3 \text{ hundreds} = 2 \text{ hundreds} + 10 \text{ tens}$.<br><br>Again, rewrite 10 tens as tens and ones, $10 \text{ tens} = 9 \text{ tens} + 10 \text{ ones}$.<br><br>Thus, $4 \text{ ones} + 10 \text{ ones} = 14 \text{ ones}$.<br><br>$$\begin{array}{r} \overset{2}{\cancel{3}}\,\overset{\overset{9}{10}}{\cancel{0}}\,\overset{14}{\cancel{4}} \\ -\ 1\ \ 4\ \ 6 \\ \hline 1\ \ 5\ \ 8 \end{array}$$ | **10.** Subtract $709 - 248$. |

| Example | Student Practice |
|---|---|

**11.** Find the perimeter of the shape consisting of rectangles.

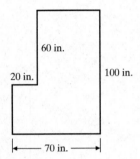

**12.** Find the perimeter of the shape consisting of rectangles.

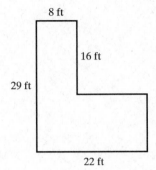

To find the perimeter we must find the distance around the figure. Therefore, we must find the measures of the unlabeled sides.

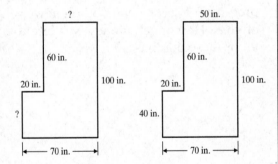

To find the unlabeled side on the top subtract $70 - 20 = 50$. Thus, the unlabeled top side is 50 inches.

To find the unlabeled left side, subtract $100 - 60 = 40$. Thus, the unlabeled left side is 40 inches.

To find the perimeter, add the lengths of the six sides.

$50 + 100 + 70 + 40 + 20 + 60 = 340$

Thus, the perimeter is 340 in.

**Extra Practice**

1. Translate the following into numbers and symbols.

   three hundred one subtracted from four hundred

2. Evaluate the expression using the given values of $x$ and $y$.

   $x - y - 99$ when $x = 700$ and $y = 101$

3. Subtract and check.

   $$\begin{array}{r} 109,000 \\ -\ \ 67,898 \\ \hline \end{array}$$

4. Barbara Ann received $175 in birthday money from her relatives. She spent $35 on a new hairstyle, $26 to fix the front tire on her bicycle, and $59 on new clothes. How much money was left after all of these purchases?

**Concept Check**

Explain why when we subtract $800 - 35$, we change 8 to 7 in the borrowing process.

Name: _____     Date: _____

Instructor: _____     Section: _____

**Chapter 1 Whole Numbers and Introduction to Algebra**
**1.4 Multiplying Whole Number Expressions**

**Vocabulary**
multiplication • array • factors • product • trailing zeros
commutative property of multiplication • multiplication property of 0
identity property of 1 • associative property if multiplication

1. The result of a multiplication problem is called a(n) _____.

2. The _____ states that when any number is multiplied by 0, the product is 0.

3. The _____ states that changing the order of factors does not change the product.

4. The numbers or variables we multiply are called _____.

| **Example** | **Student Practice** |
|---|---|
| **1.** Draw two arrays that represent the multiplication 3 times 4. | **2.** Draw two arrays that represent the multiplication 2 times 5. |
| There are two arrays consisting of twelve items that represent the multiplication 3 times 4. One array has 4 rows and 3 columns, and the other one has 3 rows and 4 columns.  | |
| **3.** Identify the product and the factors in the given equation.  $5(4) = 20$  5 and 4 are the factors and 20 is the product. | **4.** Identify the product and the factors in the given equation.  $3b = 24$ |

Vocabulary Answers: 1. product  2. multiplication property of 0  3. commutative property of multiplication
4. factors

| Example | Student Practice |
|---|---|
| **5.** Translate using numbers and symbols.<br><br>The product of four and a number<br><br>$4 \cdot n = 4n$ | **6.** Translate using numbers and symbols.<br><br>Five times a number |
| **7.** Multiply $4 \cdot 2 \cdot 4 \cdot 5$.<br><br>Use the commutative property to change the order of factors so that one factor is 10.<br><br>$4 \cdot 2 \cdot 4 \cdot 5 = (4 \cdot 4) \cdot (2 \cdot 5)$<br><br>Since $4 \cdot 4 = 16$ and $2 \cdot 5 = 10$, we have that $(4 \cdot 4) \cdot (2 \cdot 5) = 16 \cdot 10$.<br><br>To multiply 16(10), write 16 and attach a zero at the end.<br><br>$16 \cdot 10 = 160$ | **8.** Multiply $2 \cdot 3 \cdot 5 \cdot 6$. |
| **9.** Simplify $2(3)(n \cdot 7)$.<br><br>Rewrite using familiar notation, and multiply $2 \cdot 3 = 6$.<br><br>$2(3)(n \cdot 7) = 2 \cdot 3 \cdot (n \cdot 7) = 6 \cdot (n \cdot 7)$<br><br>Now, change the order of factors and regroup.<br><br>$6 \cdot (n \cdot 7) = 6 \cdot (7 \cdot n) = (6 \cdot 7) \cdot n$<br><br>Multiply and write in standard notation.<br><br>$(6 \cdot 7) \cdot n = 42 \cdot n$<br>$\qquad\qquad\quad = 42n$<br><br>Thus, $2(3)(n \cdot 7) = 42n$. | **10.** Simplify $6(7)(n \cdot 5)$. |

| Example | Student Practice |
|---|---|
| **11.** Multiply $(547)(600)$. | **12.** Multiply $(458)(300)$. |

**11.** Multiply $(547)(600)$.

Since the number 600 has trailing zeros, multiply the nonzero digits and attach the trailing zeros to the right side of the product. Bring down the trailing zeros and multiply $6(7) = 42$. Place the 2 in the hundreds place and carry the 4.

$$\begin{array}{r} \overset{4}{5}47 \\ \times\quad 600 \\ \hline 200 \end{array}$$

Next, multiply $6\cdot4$ and add the carried digit. Place the 8 in the thousands place and carry the 2. Repeat the process for $6\cdot5$ to complete the multiplication.

$$\begin{array}{r} \overset{2\,4}{5}47 \\ \times\quad 600 \\ \hline 328,200 \end{array}$$

**12.** Multiply $(458)(300)$.

---

**13.** Multiply $857(43)$.

To multiply $857(43)$, multiply $857(3+40)$ or $857(3)+857(40)$ using the condensed form. First multiply $3(857) = 2571$. Then to find the product of $40(857)$, multiply $4(857)$ and add one trailing zero. Then finally, add the partial products.

$$\begin{array}{r} 857 \\ \times\ 43 \\ \hline 2571 \\ \underline{34280} \\ 36,851 \end{array}$$

**14.** Multiply $105(3245)$.

| Example | Student Practice |
|---|---|
| **15.** Jessica drove an average speed of 60 miles per hour for 7 hours (per hour means each hour). How far did she drive? | **16.** Mike earns \$8 per hour as a retail cashier. How much will he earn if he works 38 hours? |

We draw a diagram and see that this is a situation that involves repeated addition, which indicates that we multiply.

| 60 miles | 60 miles | 60 miles | and so on.

| 60 | Miles driven each hour |
|---|---|
| $\times\ 7$ | Number of hours driven |
| 420 | Total miles driven |

Check the answer using the diagram. We see that in 3 hours, Jessica drove 180 miles $(60+60+60)$. Thus, in 6 hours she drove 360 miles $(180+180)$. Now, since she drove 60 miles the seventh hour, we add $360+60=420$ miles.

**Extra Practice**

**1.** State what property of multiplication is represented in the mathematical statement.

$$3\cdot(4\cdot2)=(3\cdot4)\cdot2$$

**2.** Translate using numbers and symbols. Do not evaluate.

Four doubled

**3.** Multiply.

$$\begin{array}{r} 12{,}402 \\ \times\ \ \ 607 \end{array}$$

**4.** Rita bought three boxes of blank CDs. If each box holds 25 CDs, how many CDs did Rita buy?

**Concept Check**

Explain what to do with the zeros when you multiply $546\times2000$.

**Chapter 1 Whole Numbers and Introduction to Algebra**
**1.5 Dividing Whole Number Expressions**

**Vocabulary**
division  •  quotient  •  divisor  •  dividend  •  undefined  •  remainder

1. Division by 0 is said to be _____.

2. The result of a division problem is called the _____.

3. The number sometimes "left over" in a division problem is called the _____.

4. The process of repeated subtraction is called _____.

| **Example** | **Student Practice** |
|---|---|
| **1.** Write the division statement that corresponds to the following situation. You need not carry out the division. | **2.** Write the division statement that corresponds to the following situation. You need not carry out the division. |
| 120 students in a band are marching in 5 rows. How many students are in each row? | Ria has a total of $180 and she would like to divide it out among 9 charities. How much money will each charity receive? |
| We draw a picture. We want to split 120 into 5 equal groups. | |

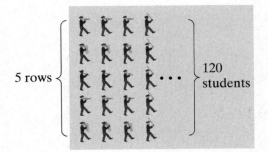

5 rows / 120 students

The division statement that corresponds to this situation is $120 \div 5$.

Vocabulary Answers: 1. undefined  2. quotient  3. remainder  4. division

| Example | Student Practice |
|---|---|
| **3.** Translate using numbers and symbols. | **4.** Translate using numbers and symbols. |
| The quotient of forty-six and two | The quotient of four and forty-eight |
| The phrase "the quotient" indicates division, and the order of the numbers indicates the dividend, 46, and the divisor, 2. | |
| $46 \div 2$ | |
| **5.** Divide $18 \div 3$. | **6.** Divide $36 \div 9$. |
| Think, $18 = \boxed{?} \cdot 3$. | |
| $18 \div 3 = \boxed{?}$ | |
| $18 = \boxed{6} \cdot 3$ | |
| Thus, $18 \div 3 = 6$. | |
| **7.** Divide. | **8.** Divide. |
| **(a)** $0 \div 9$ | **(a)** $\dfrac{28}{28}$ |
| 0 divided by any nonzero number is equal to 0. | |
| $0 \div 9 = 0$ | |
| **(b)** $9 \div 0$ | **(b)** $19 \div 0$ |
| Zero can never be the divisor in a division problem. So, $9 \div 0$ is undefined. | |
| **(c)** $\dfrac{16}{16}$ | **(c)** $0 \div 15$ |
| Any number divided by itself is 1, $16 \div 16 = 1$. | |

| Example | Student Practice |
|---|---|
| **9.** Divide and check your answer. $293 \div 41$ | **10.** Divide and check your answer. $17,985 \div 57$ |

**9.** Divide and check your answer. $293 \div 41$

First guess (too large): First guess that 41 times 8 is close to 293. Write 8 in the quotient.

$$\begin{array}{r} 8 \\ 41\overline{)293} \\ -328 \end{array}$$

Check the first guess, $41(8) = 328$; Our guess is too large, so we must adjust.

Second guess (too small): Try 6.

$$\begin{array}{r} 6 \\ 41\overline{)293} \\ -246 \\ \hline 47 \end{array}$$

Check the second guess, $41(6) = 246$; 246 is less than 293, but 47 is not less than 41. Our guess is too small so we must adjust.

Third guess: Try 7.

$$\begin{array}{r} 7\ R6 \\ 41\overline{)293} \\ -287 \\ \hline 6 \end{array}$$

Check the third guess, $41(7) = 287$; 287 is less than 293, 6 is less than 41. We do not need to adjust our guess, and 6 is the remainder.

Verify that the answer is correct by multiplying the divisor by the quotient and then adding the remainder.

19

| Example | Student Practice |
|---|---|
| **11.** Twenty-six students in Ellis High School entered their class project in a contest sponsored by the Falls City Baseball Association. The class won first place and received 250 tickets to the baseball play-offs. The teacher gave each student in the class an equal number of tickets, then donated the extra tickets to a local Boys and Girls Club. How many tickets were donated to the Boys and Girls Club? | **12.** Mike, Julie, Amber, and Ali pooled their allowances and purchased a box of 47 candy bars. They agreed to split the box evenly among the four of them and give any leftover candy to their friend Stacey. How many candy bars did Stacey receive? |

Since we must split 250 equally among 26 students, we divide.

$$\begin{array}{r} 9 \text{ R } 16 \\ 26\overline{)250} \\ \underline{234} \\ 16 \end{array}$$

Since 16 tickets are left over, 16 tickets are donated to the Boys and Girls Club.

**Extra Practice**

**1.** Divide $77 \div 0$.

**2.** Divide $21\overline{)2063}$.

**3.** Translate using numbers and symbols.

fifty-two divided by a number

**4.** Keith planted eight rows of tomatoes all containing the same number of plants. He planted 96 tomato plants all together. How many plants were in each row?

**Concept Check**
Explain the next 2 steps for this division problem.

$$\begin{array}{r} 2 \\ 13\overline{)2645} \\ \underline{26} \\ 04 \end{array}$$

**Chapter 1 Whole Numbers and Introduction to Algebra**
**1.6 Exponents and the Order of Operations**

**Vocabulary**
factor • exponent form • exponent • base • squared • cubed • order of operations

1.  The first step in the _____ is to perform operations inside parentheses.

2.  If the value of the exponent is 2, we say the base is _____.

3.  In the expression $3^5$, the number 5 is called the _____.

4.  In the expression $3^5$, the number 3 is called the _____.

| Example | Student Practice |
|---|---|
| **1.** Write in exponent form. | **2.** Write in exponent form. |
| **(a)** $2\cdot2\cdot2\cdot2\cdot2\cdot2$ | **(a)** $7\cdot7\cdot7\cdot7\cdot7\cdot7\cdot7\cdot7$ |
| $2\cdot2\cdot2\cdot2\cdot2\cdot2 = 2^6$ | |
| **(b)** $y\cdot y\cdot y\cdot3\cdot3\cdot3\cdot3$ | **(b)** $a\cdot a\cdot a\cdot a\cdot9\cdot9\cdot9$ |
| $y\cdot y\cdot y\cdot3\cdot3\cdot3\cdot3 = y^3\cdot3^4$, or $3^4 y^3$ | |
| Note that it is standard to write the number before the variable in a term. Thus $y^3 3^4$ is written $3^4 y^3$. | |
| **3.** Write as a repeated multiplication. | **4.** Write as a repeated multiplication. |
| **(a)** $n^3$ | **(a)** $y^7$ |
| $n^3 = n\cdot n\cdot n$ | |
| **(b)** $6^5$ | **(b)** $9^6$ |
| $6^5 = 6\cdot6\cdot6\cdot6\cdot6$ | |

Vocabulary Answers: 1. order of operations  2. squared  3. exponent  4. base

| Example | Student Practice |
|---|---|
| **5.** Evaluate each expression. | **6.** Evaluate each expression. |
| (a) $3^3$ | (a) $4^5$ |
| $3^3 = 3 \cdot 3 \cdot 3 = 27$ | |
| (b) $1^9$ | (b) $1^{15}$ |
| $1^9 = 1$ | |
| We do not need to write out this multiplication because repeated multiplication of 1 is equal to 1. | |
| (c) $2^4$ | (c) $10^3$ |
| $2^4 = 2 \cdot 2 \cdot 2 \cdot 2 = 16$ | |
| **7.** Evaluate $10^7$. | **8.** Evaluate $10^9$. |
| Write 1. Then, since the exponent is 7, attach 7 trailing zeros. | |
| $10^7 = 10,000,000$ | |
| **9.** Evaluate $x^3$ for $x = 3$. | **10.** Evaluate $y^4$ for $y = 5$. |
| Replace $x$ with 3. | |
| $x^3 = (3)^3$ | |
| Write as repeated multiplication, and then multiply. | |
| $(3)^3 = 3 \cdot 3 \cdot 3 = 27$ | |
| When $x = 3$, $x^3$ is equal to 27. | |

| **Example** | **Student Practice** |
|---|---|
| **11.** Translate using symbols. | **12.** Translate using symbols. |
| (a) Five cubed | (a) Twelve squared |
| Five cubed $= 5^3$ | |
| | (b) Five to the seventh power |
| (b) $y$ to the eighth power | |
| $y$ to the eighth power $= y^8$ | |

**13.** Evaluate $4 + 3\left(6 - 2^2\right) - 7$.

Always perform the calculations inside the parentheses first. Once inside the parentheses, proceed using the order of operations. Within the parentheses, exponents have the highest priority, $2^2 = 4$.

$$4 + 3\left(6 - 2^2\right) - 7 = 4 + 3(6 - 4) - 7$$

Finish all operations inside the parentheses. Subtract $6 - 4 = 2$.

$$4 + 3(6 - 4) - 7 = 4 + 3(2) - 7$$

Now the highest priority is multiplication, $3 \cdot 2 = 6$.

$$4 + 3(2) - 7 = 4 + 6 - 7$$

Next, add first, $4 + 6 = 10$.

$$4 + 6 - 7 = 10 - 7$$

Finally, subtract last, $10 - 7 = 3$.

Thus, $4 + 3(6 - 2^2) - 7 = 3$.

**14.** Evaluate $4 + 9\left(20 - 3 \cdot 5\right) - 6$.

| Example | Student Practice |
|---|---|
| **15.** Evaluate $\dfrac{(6+6\div 3)}{(5-1)}$. | **16.** Evaluate $\dfrac{(9+12\div 4)}{(8-5)}$. |

Rewrite the problem as division and then follow the order of operations.

$$(6+6\div 3)\div(5-1)$$

First, perform operations inside parentheses. $6\div 3=2$; $5-1=4$.

$$(6+6\div 3)\div(5-1)=(6+2)\div 4$$
$$=8\div 4$$

Finally, divide.

$$8\div 4=2$$

**Extra Practice**

**1.** Write in exponent form.

$a\times a\times a\times a\times a$

**2.** Translate using symbols.

$y$ to the sixth power

**3.** Using the correct order of operations, evaluate the expression.

$5\times 2^3-4\times(14\div 7)$

**4.** Using the correct order of operations, evaluate the expression.

$$\dfrac{54(13-10)-18}{3^2-3}$$

**Concept Check**

Explain in what order you would do the steps to evaluate $50+3\times 5^2\div 25$.

**Chapter 1 Whole Numbers and Introduction to Algebra**
**1.7 More on Algebraic Expressions**

**Vocabulary**
distributive property   •   parentheses   •   order of operations   •   distribute

1. It is necessary to include _____ when the phrases "sum of" or "difference of" are used or the answer obtained may be wrong.

2. The _____ states that if $a$, $b$, and $c$ are numbers or variables, then $a(b+c)=ab+ac$ and $a(b-c)=ab-ac$.

3. We _____ $a$ over addition and subtraction by multiplying every number or variable inside the parentheses by $a$.

4. When we translate phrases into numbers and symbols we must take care to preserve the _____ indicated by the phrase.

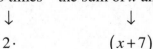

| **Example** | **Student Practice** |
|---|---|
| **1.** Translate using numbers and symbols. | **2.** Translate using numbers and symbols. |
| **(a)** Two times $x$ plus seven | **(a)** Eight times $a$ plus eleven |

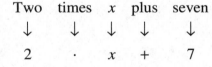

| | |
|---|---|
| **(b)** Two times the sum of $x$ and seven | **(b)** Four times the sum of $n$ and six |
| The key phrase "sum of" indicates that $x+7$ is placed in parentheses. | |

Two times    the sum of $x$ and seven
   ↓               ↓
  2·              $(x+7)$

Thus, "two times the sum of $x$ and seven" translates to $2(x+7)$.

Vocabulary Answers: 1. parentheses  2. distributive property  3. distribute  4. order of operations

| Example | Student Practice |
|---|---|
| **3.** Evaluate $\dfrac{(2a+3)}{7}$ for $a=9$. | **4.** Evaluate $\dfrac{(7y-5)}{4}$ for $y=3$. |

Replace $a$ with 9.

$$\dfrac{(2a+3)}{7}=\dfrac{(2\cdot 9+3)}{7}$$

Now, multiply.

$$\dfrac{(2\cdot 9+3)}{7}=\dfrac{(18+3)}{7}$$

Next, complete the operations within the parentheses. Then, divide.

$$\dfrac{(18+3)}{7}=\dfrac{21}{7}$$
$$=3$$

---

**5.** Evaluate $\dfrac{(x^2-2)}{y}$ for $x=4$ and $y=2$.    **6.** Evaluate $\dfrac{(a^4-9)}{b}$ for $a=3$ and $b=12$.

Replace $x$ with 4 and $y$ with 2.

$$\dfrac{(x^2-2)}{y}=\dfrac{(4^2-2)}{2}$$

Now square 4 and then subtract:
$4^2=16,\ 16-2=14$.

$$\dfrac{(4^2-2)}{2}=\dfrac{14}{2}$$

Finally, divide.

$$\dfrac{14}{2}=7$$

| Example | Student Practice |
|---|---|
| **7.** Use the distributive property to simplify. | **8.** Use the distributive property to simplify. |

**7.** Use the distributive property to simplify.

$$3(x-2)$$

First, multiply 3 times $x$, then multiply 3 times 2.

$$3(x-2)=3\cdot x-3\cdot 2$$

Finally, simplify your answer.

$$3\cdot x-3\cdot 2=3x-6$$

Thus, $3(x-2)=3x-6$.

**8.** Use the distributive property to simplify.

$$4(8+y)$$

---

**9.** Simplify.

$$2(y+1)+4$$

First use the distributive property to multiply $2(y+1)$ and then simplify.

$$2(y+1)+4=2\cdot y+2\cdot 1+4$$
$$=2y+2+4$$

Now, simplify the resulting expression.

$$2y+2+4=2y+6$$

Thus, $2(y+1)+4=2y+6$.

**10.** Simplify.

$$9(y+4)+5$$

**Extra Practice**

1. Translate using numbers and symbols.

   Eight times the difference of $x$ and one

2. Evaluate for the given values.

   $$\frac{\left(m^2 - 5\right)}{4} \text{ when } m = 5$$

3. Evaluate for the given values.

   $3mn + 2m + 5n$ when $m = 4$ and $n = 8$

4. Use the distributive property to simplify.

   $3(x + y + 4) - 2(y + 1)$

**Concept Check**

Simplify $5(x+1)$, then evaluate $5(x+1)$ for $x = 2$. Compare results and state the difference in the process to simplify and to evaluate.

**Chapter 1 Whole Numbers and Introduction to Algebra**
**1.8 Introduction to Solving Linear Equations**

**Vocabulary**

term • coefficient • constant term • variable term • like terms
expression • equation • solution • solve • combine like terms

1. Terms that have identical variable parts are called _____.

2. In a(n) _____, we use an equals sign $(=)$ to indicate that two expressions are equal in value.

3. The numerical part of a term is called the _____ of the term.

4. A(n) _____ is a number, a variable, or a product of a number and one or more variables.

| **Example** | **Student Practice** |
|---|---|
| **1.** Write a term that represents the following.<br><br>Two $y$'s<br><br>Two $y$'s $= 2y$ | **2.** Write a term that represents the following.<br><br>$b + b + b + b + b + b$ |
| **3.** Identify like terms, then combine like terms. $4xy + 8y + 2xy$<br><br>Identify and group like terms.<br><br>$4xy + 8y + 2xy = (4xy + 2xy) + 8y$<br><br>Add the numerical coefficients of like terms.<br><br>$(4xy + 2xy) + 8y = (4 + 2)xy + 8y$<br><br>Thus, $4xy + 8y + 2xy = 6xy + 8y$. We cannot combine $8y$ with $6xy$ since the variable parts are not the same. | **4.** Identify like terms, then combine like terms. $8y + 2x + 7y + 9x$ |

Vocabulary Answers: 1. like terms  2. equation  3. coefficient  4. term

| Example | Student Practice |
|---|---|
| **5.** Write the perimeter of the rectangular field as an algebraic expression and simplify. | **6.** Write the perimeter of the triangular figure as an algebraic expression and simplify. |

**5.** Write the perimeter of the rectangular field as an algebraic expression and simplify.

4a + 7b

2a + 3b

Since the figure is a rectangle, opposite sides are equal.

4a + 7b

2a + 3b          2a + 3b

4a + 7b

We add all sides to find the perimeter.

$$(2a+3b)+(4a+7b)+(2a+3b)+$$
$$(4a+7b)$$

Now use the associative and commutative properties to change the order of addition and regroup. Then finally, combined like terms.

$$(2a+2a+4a+4a)+(3b+3b+7b+7b)$$
$$=12a+20b$$

Thus, the algebraic expression for the perimeter is $12a+20b$.

**6.** Write the perimeter of the triangular figure as an algebraic expression and simplify.

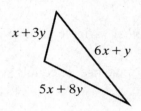

$x + 3y$

$6x + y$

$5x + 8y$

---

**7.** Is 2 a solution to $6-x=9$?

If 2 is a solutions to $6-x=9$, when we replace $x$ with the value $x=2$ we will get a true statement.

Replace the variable with 2 and simplify.

$$6-2\overset{?}{=}9$$

$$4\overset{?}{=}9$$

This is a false statement. Since $4=9$ is not a true statement, 2 is not a solution to $6-x=9$.

**8.** Is 7 a solution to $x+5=14$?

30

| Example | Student Practice |
|---|---|
| **9.** Solve the equation $3+n=10$ and check your solution. | **10.** Solve the equation $n+6=11$ and check your solution. |

To solve the equation $3+n=10$, answer the question: "Three plus what number is equal to ten?" Using addition facts we see that the answer, or solution, is 7. To check the solution, we replace $n$ with the value $n=7$ and verify that we get a true statement.

$$3+n \ = \ 10$$

$$3+7 \ \overset{?}{=} \ 10$$

$$10 \ = \ 10 \ \text{True}$$

Since we get a true statement, the solution to $3+n=10$ is 7 and is written $n=7$.

| **11.** Simplify using the associative and commutative properties and then find the solution to the equation $(5+n)+1=9$. | **12.** Simplify using the associative and commutative properties and then find the solution to the equation $(6+a)+3=13$. |

First, apply the commutative property. Next, apply the associative property and simplify.

$$(n+5)+1=9$$

$$n+(5+1)=9$$

$$n+6=9$$

Now, solve $n+6=9$. Answer the question "What number plus 6 is equal to 9?"

$$n=3$$

Thus, the solution to the equation is 3.

31

| Example | Student Practice |
|---|---|
| **13.** Translate, then solve. Double what number is equal to eighteen? | **14.** Translate, then solve. Triple what number is equal to twenty-one? |

"Double" translates to "2 times" or "$2 \cdot$."

"What number" translates to "$n$."

"is equal to" translates to "$=$."

Thus, the equation is $2 \cdot n = 18$. Use multiplication facts to find $n$.

Since $2 \cdot 9 = 18$, $n = 9$.

**Extra Practice**

**1.** Combine like terms.

$14ab + 8 + 4ab + 3$

**2.** Translate the mathematical symbols using words.

$x + 8 = 12$

**3.** Solve and check your answer.

$(a + 5) + 2 = 16$

**4.** Translate to an equation. Do not solve the equation.

Three plus what number equals twelve?

**Concept Check**
Explain the difference in the process you must use to complete **(a)** and **(b)**.
**(a)** Combine like terms. $3x + x + 2x$
**(b)** Solve. $3x + x + 2x = 12$

**Chapter 1 Whole Numbers and Introduction to Algebra**
**1.9 Solving Applied Problems Using Several Operations**

**Vocabulary**

estimate   •   understand the problem   •   calculate and state the answer   •   check the answer

1.  The first step in solving applied problems is to _____, which involves reading the problem and organizing the information.

2.  To _____ a sum or difference, round each number to the same round-off place and then find the sum or difference.

| **Example** | **Student Practice** |
|---|---|
| 1.  Some sample sale prices for 2010 Ford motor vehicles are listed below. | 2.  Use the figure in example **1** to answer the following. Estimate the difference in the price between an F-150 SVT Raptor and a Focus SE Coupe by rounding each price to the nearest thousand. |

**Manufacturers Suggested Retail Prices on 2010 Ford Vehicles**

| | |
|---|---|
| F-150 SVT Raptor | $38,020 |
| Taurus SEL | $28,195 |
| Focus SE Coupe | $15,895 |
| Mustang V6 Premium Coupe | $26,695 |

*Source:* www.fordvehicles.com

Estimate the difference in the price between an F-150 SVT Raptor and a Mustang V6 Premium Coupe by rounding each price to the nearest thousand.

The exact value for the price of the F-150 SVT Raptor is 38,020 and the rounded value is 38,000. The exact value for the price of the Mustang V6 Premium Coupe is 26, 695 and the rounded value is 27,000. To estimate the difference in the cost of the two vehicles, subtract the two rounded figures, $38,000 - 27,000 = 11,000$. Thus, the estimated difference in price is $11,000.

Vocabulary Answers: 1. understand the problem  2. estimate

| Example | Student Practice |
|---|---|
| **3.** A frequent-flyer program offered by many major airlines to first-class passengers awards 3 frequent-flyers mileage points for every 2 miles flown. When customers accumulate a certain number of frequent-flyer points, they can cash them in for free air travel, ticket upgrades, or other awards. How many frequent-flyer points would a customer accumulate after flying 3500 miles in first class? | **4.** A frequent-flyer program offered by many major airlines to first-class passengers awards 3 frequent-flyers mileage points for every 2 miles flown. When customers accumulate a certain number of frequent-flyer points, they can cash them in for free air travel, ticket upgrades, or other awards. How many frequent-flyer points would a customer accumulate after flying 5500 miles in first class? |

Sometimes, drawing charts or pictures can help us understand the problem as well as plan our approach to solving the problem.

$$2 \text{ miles} + 2 \text{ miles} \cdots = 3500 \text{ miles}$$

$$3 \text{ points} + 3 \text{ points} \cdots = ? \text{ points}$$

Organize the information and make a plan using the Mathematics Blueprint.

We divide to find how many groups of 2 are in 3500, $3500 \div 2 = 1750$.

Then, we multiply 1750 times 3 to find the total points earned, $1750 \cdot 3 = 5250$ points.

Check: If the customer earned 4 points (instead of 3) for every 2 miles traveled, we could just double the mileage to find the points earned, 2 miles earns 4 points and 3500 miles earns 7000 points. Since the customer earned a little less than 4 points, the total should be less than 7000. It is: $5250 < 7000$. The customer also earned more points than miles traveled, so the total points should be more than the total miles traveled. It is: $5250 > 3500$. Our answer is reasonable.

| Example | Student Practice |
|---|---|

**5.** Koursh was offered two different jobs: a 40-hour-a-week store assistant management position that pays $14 per hour and an executive secretary position paying a monthly salary of $2600. Which job pays more per year?

Read the problem carefully and create a Mathematics Blueprint. We write out a process to find the answer.

$$\$14 \times 40 = \$560 \qquad \text{Pay for 1 week (management)}$$

$$\$560 \times 52 = \$29,120 \qquad \text{Pay for 1 year (management)}$$

$$\$2600 \times 12 = \$31,200 \qquad \text{Pay for 1 year (secretary)}$$

The yearly pay is $29,120 for the management position and $31,200 for the secretary's position.

The secretary's position pays more per year.

Check: We estimate the assistant manager's pay per year by rounding $14 per hour to $10 and 52 weeks to 50 weeks.

$10 \times 40$ hr $= \$400$ per week;
$400 \times 50$ weeks $= \$20,000$ per year

We estimate the secretary's pay per year by rounding 12 months per year to 10.

$10 \times \$2600 = \$26,000$ per year

Since $\$26,000 > \$20,000$, the secretary position pays more.

**6.** Anny works 40 hours a week and earns $17 an hour as a payroll clerk. She is considering accepting a job offer to work as an assistant office manager earning a salary of $3500 per month. Which job pays more per year?

**Extra Practice**

1. Karen wants to buy the following four items shown in the chart. Round the price of each item to the nearest hundred dollars.

| Item | Price |
|---|---|
| Bedspread | $189 |
| Digital Camera | $331 |
| Fountain | $94 |
| Stained Glass | $453 |

2. Using the chart and rounded values from extra practice **1**, estimate the total cost of all four items.

3. Last week, Maureen worked 53 hours. She earns $10 per hour for the first 40 hours worked, and then earns $15 per hour every hour after 40 hours worked. How much money should Maureen expect to be paid for last week's work?

4. A textbook author estimates that she can write 3 pages per day. How many pages can she expect to write in 30 days? 60 days?

**Concept Check**

At the end of January, Sahara had $200 left in her vacation savings account and $1000 in her household savings account. Each month for the next six months, Sahara plans to put $100 in her vacation account and $200 in her household account. In addition, she plans to split her $900 tax return equally between both accounts. Explain how to determine if Sahara will have enough money in her vacation account at the end of six months to take a $1500 vacation.

# MATH COACH

*Mastering the skills you need to do well on the test.*

Watch the **MATH COACH** videos in MyMathLab®or on You Tube™  while you work the problems below. These helpful hints will help you avoid making common errors on test problems.

### Subtract Whole Numbers with Borrowing—Problem 7(b)

Subtract.  $\begin{array}{r} 20{,}105 \\ -7{,}826 \\ \hline \end{array}$

> **Helpful Hint:** It is wise to show the borrowing steps. This will help you avoid a borrowing error.

Look at your work for Problem 7(b). Examine your steps. Do your borrowing steps match the solution below?
Yes _____   No _____

If you answered No, be sure to write your borrowing steps carefully and then check each subtraction step for errors. Stop now and rework the problem.

If you answered Yes, and still got an incorrect answer, check each subtraction step for errors.

Write out the borrowing steps:

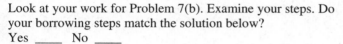

If you answered Problem 7(b) incorrectly, go back and rework the problem using these suggestions.

---

### Performing Long Division with Whole numbers—Problem 11(b) Divide $5523 \div 46$.

> **Helpful Hint:** When performing long division, be sure to line up each column exactly as shown below. Accuracy in alignment will increase your chance of completing the problem correctly. Remember that a remainder results when there are no more numbers to bring down from the dividend.

Look at your work from Problem 11(b). Compare each line of your work with the problem below.

Did you align numbers correctly?
Yes _____   No _____

Did you obtain the correct value each time you subtracted?
Yes _____   No _____

$$\begin{array}{r} 46{\overline{\smash{\big)}\,5523}} \\ \underline{46\phantom{00}} \\ 92\phantom{0} \\ \underline{92\phantom{0}} \\ 03 \end{array}$$

If you answered No to either question, stop and rework the problem now.

Did you get 12 R3 for your answer?

Yes _____   No _____

If you answered Yes, then you stopped dividing too soon. Because 46 cannot be divided into 3, we must write 0 in the quotient and continue to divide.

Now go back and rework the problem using these suggestions.

## Evaluate Algebraic Expression—Problem 16(a)

Evaluate $2x - 3y$ if $x$ is equal to 16 and $y$ is equal to 4.

As your first step, did you replace $2\boxed{x} - 3\boxed{y}$ with $2 \cdot \boxed{16} - 3 \cdot \boxed{4}$ ? Yes _____ No _____

If you answered No, stop and complete this step.

Did you multiply $2 \cdot \boxed{16}$ for your next step?
Yes _____ No _____

If you answered No, then you *did not* follow the proper order of operations. Remember to complete all multiplications before you subtract.

If you answered Problem 11(b) incorrectly, go back and rework the problem using these suggestions.

---

## Solving Applied Problems Using Several Operations—Problem 26(a) and (b)

Tickets to a play were $25 for adults and $18 for children. 412 adult tickets were sold and 280 child tickets were sold.
  (a)  Find the total income from the sale of tickets.
  (b)  If the expenses for the play were $7350, how much profit was made?

Reread Problem 26 carefully, then fill in the information needed to solve the problem.

**What are the facts?** There are _____ adult tickets. They cost $_____ each. There are _____ child tickets. They cost $_____ each.

**What am I asked to do for Part (a)?** You must find the total cost of tickets (income).

**What type of calculation must I perform?** Did you multiply $412 \times \$25$ to find the income from adult tickets and $280 \times \$18$ to find the income from child tickets? Then did you add these results? Yes _____ No _____

If you answered No, stop now and use these hints to complete Part (a) correctly.

**What am I asked to do for Part (b)?**
You must find the profit if the expenses were $7350.

**What type of calculation must I perform?**
Did you calculate
$\text{Income} - \text{Expenses} = \text{Profit}$ ?
Yes _____ No _____

If you answered No, stop now and use these hints to complete Part (b) correctly.

Try to write out all the facts of the situation to help you keep track of the details and determine what type of calculations are needed to solve the problem.

Now go back and rework the problem using these suggestions.

Name: _____     Date: _____

Instructor: _____     Section: _____

**Chapter 2 Integers**
**2.1 Understanding Integers**

**Vocabulary**
negative numbers   •   positive numbers   •   number line   •   signed numbers
integers   •   opposites   •   absolute value

1.  Positive numbers are to the right of 0 and negative numbers are to the left of 0 on the
    _____.

2.  Signed numbers, …, $-3$, $-2$, $-1$, 0, 1, 2, …, are also called _____.

3.  Numbers that are the same distance from zero but lie on the opposite sides of zero on the
    number line are called _____.

4.  Numbers that are less than 0 are called _____.

| **Example** | **Student Practice** |
|---|---|
| **1.** Graph $-5$, $-3$, 1, and 5 on a number line. <br><br> We draw a dot in each of the correct locations on the number line. We start at zero for each and count five places in the negative direction for $-5$, three places in the negative direction for $-3$, one place in the positive direction for 1, and five places in the positive direction for 5. <br><br> 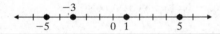 | **2.** Graph $-3, -1$, 0, and 2 on a number line. <br><br> 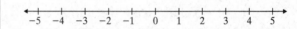 |
| **3.** Replace each ? with the inequality symbol $<$ or $>$. <br><br> **(a)** $-3$ ? $-1$ <br><br> $-3 < -1$ <br><br> **(b)** 4 ? $-5$ <br><br> $4 > -5$ | **4.** Replace each ? with the inequality symbol $<$ or $>$. <br><br> **(a)** $-6$ ? $-3$ <br><br><br> **(b)** $-5$ ? 2 |

Vocabulary Answers: 1. number line  2. integers  3. opposites  4. negative numbers

| Example | Student Practice |
|---|---|

**5.** Fill in each blank with the appropriate symbol, + or −, to describe either an increase or a decrease.

   **(a)** A discount of $5 : ___$5

   A discount of $5 results in the price decreasing: −$5

   **(b)** The temperature rises 10°F : ___10°F

   The temperature rises, or increases, by 10°F : +10°F

**6.** Fill in each blank with the appropriate symbol, + or −, to describe either an increase or a decrease.

   **(a)** A rent increase of $50: ___$50

   **(b)** A rock falling 30 ft: ___ 30 ft

---

**7.** Label −5 and the opposite of −5 on a number line.

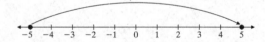

We start at −5 and locate the number that is the same distance from zero but lies on the opposite side of zero.

Thus, the opposite of −5 is 5.

**8.** Label −2 and the opposite of −2 on a number line.

---

**9.** Evaluate $-(-x)$ for $x = -9$.

   To avoid errors involving negative signs, we can place parentheses around the variables and their replacements.

   $$-(-x) = -(-(x))$$
   $$= -(-(-9))$$
   $$= -(9)$$
   $$= -9$$

**10.** Evaluate $-\left(-(-b)\right)$ for $b = -2$.

| Example | Student Practice |
|---|---|
| **11.** Replace the ? with the symbol $<$, $>$, or $=$. $\|-15\|$ ? $\|6\|$<br><br>$\|-15\|$ ? $\|6\|$<br>$\quad$ 15 ? 6<br>$\quad$ 15 $>$ 6<br>$\|-15\|$ $>$ $\|6\|$<br><br>Note that when we say $-15$ has a larger absolute value, we mean that $-15$ is a greater distance from 0 than 6 is. | **12.** Replace the ? with the symbol $<$, $>$, or $=$. $\|-105\|$ ? $\|5\|$ |
| **13.** Simplify.<br><br>$-\|-7\|$<br><br>We must find the opposite of the absolute value of $-7$.<br><br>$-\|-7\| = -(7)$<br>$\qquad\quad = -7$ | **14.** Simplify.<br><br>$-\|-(-4)\|$ |
| **15.** The line graph below indicates the low temperatures for selected cities on a typical winter day.<br><br>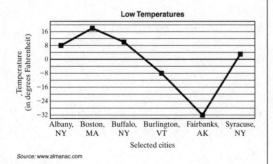<br><br>*Source:* www.almanac.com<br><br>In which city was the temperature colder, Syracuse or Burlington?<br><br>The low temperature was $3°F$ in Syracuse and $-8°F$ in Burlington. It was colder in Burlington. | **16.** Use the line graph in example **15** to answer the following.<br><br>**(a)** In which city was the temperature colder, Boston or Albany?<br><br><br>**(b)** Which three cities recorded the highest temperatures for the day? |

41

**Extra Practice**

1. Graph the following values on a number line.

   $-5, -2, 0, 3, 4$

   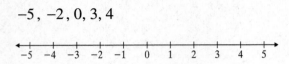

2. State both the opposite and the absolute value.

   $-21$

3. Evaluate $-\left(-(-x)\right) - \left(-|x|\right)$ for $x = -21$.

4. Which number has the greater opposite, $-4$ or $5$? Explain your thinking.

**Concept Check**

Explain how to rearrange the following numbers in order from smallest to largest:
$-1, -6, -4, -10, \ 0$

Name: _____     Date: _____

Instructor: _____     Section: _____

## Chapter 2 Integers
## 2.2 Adding Integers

**Vocabulary**
additive inverse property   •   additive inverses   •   the same sign   •   different signs

1.  3 and $-3$ are called _____.

2.  The _____ states that for any number $a$, $a+(-a)=0$ and $(-a)+a=0$. The sum of any number and its opposite is 0.

3.  Adding numbers with _____ involves finding the difference between the larger absolute value and the smaller absolute value.

4.  Adding numbers with _____ involves adding the absolute values.

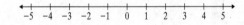

| Example | Student Practice |
|---|---|
| **1.** Answer parts **(a)** through **(c)**. | **2.** Answer parts **(a)** and **(c)**. |

**(a)** Begin at 0 on the number line and move 3 units to the left followed by another 2 units to the left.

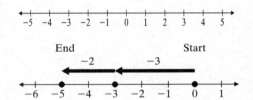

**(a)** Begin at 0 on the number line and move 2 units to the left followed by another 1 unit to the left.

**(b)** Write the math symbols that represent the situation.

$$-3+(-2)$$

**(b)** Write the math symbols that represent the situation.

**(c)** Use the number line to find the sum.

We end at $-5$, which is the sum.

$$-3+(-2)=-5$$

**(c)** Use the number line to find the sum.

Vocabulary Answers: 1. additive inverses  2. additive inverse property  3. different signs  4. the same sign

| Example | Student Practice |
|---|---|
| **3.** Add. $-1+(-3)$ <br><br> We are adding two numbers with the same sign, so keep the common sign and then add the absolute values. <br><br> $-1+(-3)=-$ <br> $-1+(-3)=-4$ | **4.** Add. $-5+(-6)$ |
| **5.** One night the temperature on Long Island, New York, dropped to $-25°F$. At dawn the temperature had risen $10°F$. Find the sum. <br><br> Use the thermometer to help determine the sum. <br><br> 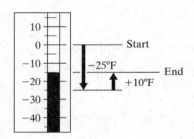 <br><br> The final temperature is $-15°F$, which is the sum: $-25°F+10°F=-15°F$. | **6.** The temperature was $25°F$ below zero. At dawn it had risen $18°F$. Find the sum. <br><br> 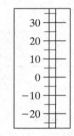 |
| **7.** Add. $2+(-3)$ <br><br> We are adding numbers with different signs, so we keep the sign of the larger absolute value and subtract. <br><br> $2+(-3)=-$ <br> $2+(-3)=-1$ | **8.** Add. $-6+8$ |
| **9.** Find $x$. $x+21=0$ <br><br> The sum of additive inverses is 0. Thus if $x+21=0$, then $x=-21$ since $-21+21=0$. | **10.** Find $x$. $-18+x=0$ |

| Example | Student Practice |
|---|---|
| **11.** Add. | **12.** Add. |

**11.** Add.

$$-3 + 9 + (-4) + 12$$

Change the order of addition and regroup.

$$[(-3) + (-4)] + (9 + 12)$$

Then, add the negative numbers.

$$[(-3) + (-4)] + (9 + 12) = -7 + (9 + 12)$$

Now, add the positive numbers.

$$-7 + (9 + 12) = -7 + 21$$

Finally, add the result.

$$-7 + 21 = 14$$

**12.** Add.

$$8 + (-4) + 3 + (-10)$$

---

**13.** Evaluate.

$$-7 + a + b \text{ for } a = -3 \text{ and } b = 9$$

We place the parentheses around the variables, and then replace each variable with the appropriate values.

$$-7 + (a) + (b) = -7 + (-3) + (9)$$

Then, add the negative numbers.

$$-7 + (-3) + (9) = -10 + 9$$

Finally, add the results.

$$-10 + 9 = -1$$

**14.** Evaluate.

$$x + (-4) + y \text{ for } x = 6 \text{ and } y = -1$$

| Example | Student Practice |
|---|---|
| **15.** Information on Micro Firm Computer Sales' profit and loss situation is given on the graph. What was the company's overall profit or loss at the end of the third quarter? | **16.** Use the bar graph in Example 15. What was Micro Firm Computer Sales' overall profit or loss at the end of the fourth quarter? |

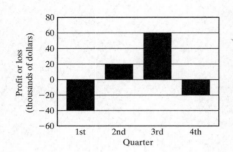

1st quarter loss + 2nd quarter profit + 3rd quarter profit = net profit

$-\$40,000 + \$20,000 + \$60,000$

$= \$40,000$

At the end of the third quarter the company had a net profit of $40,000.

**Extra Practice**

**1.** Add using the rules for addition of integers. $63 + (-21)$

**2.** Add using the rules for addition of integers. $(-15) + (-20) + (-6) + 40$

**3.** Evaluate $-20 + (-x) + (-1)$ for $x = -15$.

**4.** On Friday morning, Kevin began his camping trip at an altitude of 400 feet. He descended 350 feet on Friday, descended another 75 feet on Saturday, and then climbed up 200 feet on Sunday. Represent his altitude on Sunday night as an integer.

**Concept Check**

Without completing the calculations, explain how you can determine whether the answer is a positive or negative number. Evaluate $132 + x + y + z$ for $x = -1$, $y = -3$, and $z = -2$.

**Chapter 2 Integers**
**2.3 Subtracting Integers**

**Vocabulary**
subtraction      •      addition      •      difference      •      opposite

1.  When solving subtraction problems, the operation is usually changed to _____.

2.  When subtracting a negative number from another number, the _____ is larger.

3.  _____ of integers is not commutative or associative.

4.  To subtract, add the _____ of the second number to the first.

| Example | Student Practice |
|---|---|
| **1.** Subtract. | **2.** Subtract. |
| **(a)** $\$15 - \$20$ | **(a)** $\$25 - \$50$ |
| If we have \$15 and want to spend \$20, we are short \$5, or $-\$5$. | |
| $\$15 - \$20 = -\$5$ | |
| **(b)** $3 - 4$ | **(b)** $4 - 8$ |
| If we have 3 items and try to take 4 items, we are short 1 item, or $-1$. | |
| $3 - 4 = -1$ | |
| **3.** Rewrite each subtraction as addition of the opposite. | **4.** Rewrite each subtraction as addition of the opposite. |
| **(a)** $40 - 10 = 30$ | **(a)** $50 - 25 = 25$ |
| $40 - 10 = 30 \ \rightarrow \ 40 + (-10) = 30$ | |
| **(b)** $6 - 2 = 4$ | **(b)** $8 - 5 = 3$ |
| $6 - 2 = 4 \ \rightarrow \ 6 + (-2) = 4$ | |

Vocabulary Answers: 1. addition  2. difference  3. subtraction  4. opposite

| Example | Student Practice |
|---|---|
| **5.** Subtract. | **6.** Subtract. |
| **(a)** $-8-3$ | **(a)** $-2-3$ |
| We replace the second number by its opposite and then add using the rules for adding numbers with the same sign. | |
| $-8-3=-8+(-3)$ <br> $\qquad = -11$ | |
| **(b)** $-6-(-4)$ | **(b)** $-7-(-1)$ |
| We replace the second number by its opposite and then add using the rules for adding numbers with different signs. | |
| $-6-(-4)=-6+4$ <br> $\qquad = -2$ | |
| **7.** Subtract. | **8.** Subtract. |
| **(a)** $8-9$ | **(a)** $4-6$ |
| $8-9=8+(-9)$ <br> $\qquad = -1$ | |
| **(b)** $-3-16$ | **(b)** $-6-12$ |
| $-3-16=-3+(-16)$ <br> $\qquad = -19$ | |
| **(c)** $5-(-4)$ | **(c)** $7-(-3)$ |
| $5-(-4)=5+4$ <br> $\qquad = 9$ | |

| Example | Student Practice |
|---|---|
| **9.** Perform the necessary operations.<br><br>$4-7-5-3$<br><br>First, write all subtraction as addition of the opposite, and then group like signs.<br><br>$$\begin{aligned}4-7-5-3&=4+(-7)+(-5)+(-3)\\&=4+\left[(-7)+(-5)+(-3)\right]\end{aligned}$$<br><br>Then add all like signs before adding unlike signs.<br><br>$$\begin{aligned}4+\left[(-7)+(-5)+(-3)\right]&=4+(-15)\\&=-11\end{aligned}$$ | **10.** Perform the necessary operations.<br><br>$5-7-2-9$ |
| **11.** Evaluate $-x-y-4$ for $x=-3$ and $y=-1$.<br><br>Place parentheses around the variables and replace $x$ with $-3$ and $y$ with $-1$.<br><br>$$-(x)-(y)-4=-(-3)-(-1)-4$$<br><br>Then simplify $-(-3)$ and change each subtraction to addition of the opposite.<br><br>$$\begin{aligned}-(-3)-(-1)-4&=3-(-1)-4\\&=3+1+(-4)\end{aligned}$$<br><br>Finally, add.<br><br>$$\begin{aligned}3+1+(-4)&=4+(-4)\\&=0\end{aligned}$$ | **12.** Evaluate $-p-q-6$ for $p=-5$ and $q=-8$. |

| Example | Student Practice |
|---|---|

**13.** A portion of the Dead Sea is 1286 feet below sea level. What is the difference in altitude between Mount Carmel in Israel, which has an altitude of 1791 feet, and the Dead Sea?

Mount Carmel — 1791 ft
Sea level — — — —
Difference in altitude is 3077 ft.
Dead Sea
1286 ft below sea level or 1286 ft

We want to find the difference, so we must subtract.

$$1791 \text{ ft} - (-1286 \text{ ft}) = 1791 \text{ ft} + 1286 \text{ ft}$$
$$= 3077 \text{ ft}$$

The difference in altitude is 3077 ft.

**14.** Find the difference in altitude between a mountain 2986 feet high and a desert valley 734 feet below see level.

## Extra Practice

**1.** Subtract. $-12 - (-20)$

**2.** Subtract. $-7 + (-11) - (-15)$

**3.** Evaluate $-5 - (-x) + (-4) - y - 7$ for $x = 1$ and $y = -1$.

**4.** Martha left New York by plane at 2:00 P.M. Her flight to San Francisco took six hours, but the local time in San Francisco is three hours earlier than in New York. What was the local time in San Francisco when Martha's plane landed?

## Concept Check

Is the following problem completed correctly? Why or why not?

$$-6 - (-3) + (-7) = -6 - (-10)$$
$$= -6 + 10$$
$$= 4$$

**Chapter 2 Integers**
**2.4 Multiplying and Dividing Integers**

**Vocabulary**
odd numbers  •  even number  •  positive  •  negative  •  exponent

1. To find each consecutive _____, add 2 to the previous number. The number 0 is the first of these.

2. When multiplying or dividing, the answer will be _____ if the problem has an even number of negative signs.

3. Whole numbers that are not even numbers are _____.

4. When multiplying or dividing, the answer will be _____ if the problem has an odd number of negative signs.

| Example | Student Practice |
|---|---|
| **1.** Find the product by writing as repeated addition. $2(-3)$<br><br>$2(-3) = -3 + (-3) = -6$ | **2.** Find the product by writing as repeated addition. $4(-8)$ |
| **3.** Multiply.<br><br>**(a)** $6(7)$<br><br>The number of negative signs, 0, is even. Thus, the answer is positive.<br><br>$6(7) = +42$<br><br>**(b)** $6(-7)$<br><br>The number of negative signs, 1, is odd. Thus, the answer is negative.<br><br>$6(-7) = -$<br>$\quad\quad = -42$ | **4.** Multiply.<br><br>**(a)** $-3(-5)$<br><br><br><br>**(b)** $-3(5)$ |

Vocabulary Answers: 1. even numbers  2. positive  3. odd numbers  4. negative.

| Example | Student Practice |
|---|---|
| **5.** Multiply $(-3)(2)(-1)(4)(-3)$. | **6.** Multiply $4(-3)(-2)(-5)(-2)$. |
| The answer is negative since there are 3 negative signs and 3 is an odd number. Multiply the absolute values. | |
| $(-3)(2)(-1)(4)(-3)$ $=-\left[3(2)(1)(4)(3)\right]$ $=-72$ | |
| **7.** Evaluate $(-4)^3$. | **8.** Evaluate $(-2)^2$. |
| The answer is negative since the exponent, 3, is odd. | |
| $(-4)^3 = (-4)(-4)(-4)$ $\quad\quad = -64$ | |
| **9.** Evaluate. | **10.** Evaluate. |
| **(a)** $-3^2$ | **(a)** $-4^2$ |
| $-3^2$, "the opposite of three squared" | |
| The base is 3; we use 3 as the factor for repeated multiplication and take the opposite of the product. | |
| $-3^2 = -(3\cdot 3) = -9$ | |
| **(b)** $(-3)^2$ | **(b)** $(-4)^2$ |
| $(-3)^2$, "negative three squared" | |
| The base is $-3$; we use $-3$ as the factor for repeated multiplication. | |
| $(-3)^2 = (-3)(-3) = 9$ | |

| Example | Student Practice |
|---|---|
| **11.** Divide. | **12.** Divide. |
| (a) $36 \div 6$ | (a) $16 \div (-4)$ |
| $36 \div 6 = 6$ | |
| (b) $36 \div (-6)$ | (b) $-16 \div (-4)$ |
| $36 \div (-6) = -6$ | |
| (c) $-36 \div (-6)$ | (c) $-16 \div 4$ |
| $-36 \div (-6) = 6$ | |
| **13.** Perform the operation indicated. $\dfrac{-72}{-8}$ | **14.** Perform the operation indicated. $-15(-4)$ |
| $\dfrac{-72}{-8} = -72 \div (-8)$ $= 9$ | |
| **15.** Evaluate. | **16.** Evaluate. |
| (a) $\dfrac{m}{-n}$ for $m = -16$ and $n = -2$ | (a) $\dfrac{-c}{d}$ for $c = -64$ and $d = -8$ |
| We place parentheses around the variables and then we replace $m$ with $-16$ and $n$ with $-2$. | |
| $\dfrac{(m)}{-(n)} = \dfrac{(-16)}{-(-2)} = \dfrac{-16}{2} = -8$ | |
| (b) $x^4$ for $x = -2$ | (b) $z^7$ for $z = -1$ |
| We place parentheses around the variable and then we replace $x$ with $-2$. | |
| $(x)^4 = (-2)^4 = 16$ | |

**Extra Practice**

1. Find the product by writing as repeated addition.

   $4(-7)$

2. Perform the operation indicated.

   $(-3)^2$

3. Perform the operation indicated.

   $-100 \div 100$

4. Evaluate $(-x)^2 \div y$ for $x = 4$ and $y = -2$.

**Concept Check**

When we evaluate $(-x)(y)$ for $x = -6$ and $y = 2$, we obtain a positive number. Is this true? Why or why not?

**Chapter 2 Integers**

**2.5 The Order of Operations and Applications Involving Integers**

**Vocabulary**

replace     •     calculate     •     order of operations     •     identify

1.  When solving problems with more than one operation, you must first _____ the operation with the highest priority.

2.  Once the operation with the highest priority is known, _____ the result of this operation.

3.  Use the result of the operation to _____ the original expression of this operation.

4.  When there is more than one operation in a problem, we must follow the _____.

| **Example** | **Student Practice** |
|---|---|
| **1.** Simplify $12 - 30 \div 5(-3)^2 - 2$. | **2.** Simplify $-7 + (-6)^2 \div 9 + 3$. |

**Example** (continued):

Identify that the highest priority is exponents, $(-3)^2$. Calculate, $(-3)^2 = 9$. Replace $(-3)^2$ with 9.

$$12 - 30 \div 5(-3)^2 - 2 = 12 - 30 \div 5(9) - 2$$

Identify that the next highest priority is division. Calculate, $30 \div 5 = 6$. Replace $30 \div 5$ with 6.

$$12 - 30 \div 5(9) - 2 = 12 - 6(9) - 2$$

Continue to repeat this process, following the order of operations, to complete the simplification.

$$12 - 6(9) - 2 = 12 - 54 - 2$$
$$= 12 + (-54) + (-2)$$
$$= -44$$

Vocabulary Answers: 1. identify  2. calculate  3. replace  4. order of operations.

| Example | Student Practice |
|---|---|

**3.** Simplify $\dfrac{\left[-15+5(-3)\right]}{(13-18)}$.

**4.** Simplify $\dfrac{\left[3^4+5(-4)+3\right]}{(2^2+4)}$.

We perform the operations inside the parentheses and brackets first. We multiply; $5(-3)=-15$.

$$\frac{\left[-15+5(-3)\right]}{(13-18)}=\frac{\left[-15+(-15)\right]}{(13-18)}$$

Then, we add, $-15+(-15)=-30$.

$$\frac{\left[-15+(-15)\right]}{(13-18)}=\frac{-30}{(13-18)}$$

We subtract; $13-18=-5$.

$$\frac{-30}{(13-18)}=\frac{-30}{-5}$$

We divide last; $-30\div(-5)=6$.

$$\frac{-30}{-5}=6$$

---

**5.** Simplify $-24\div\left\{-3\cdot\left[4\div(-2)\right]\right\}$.

**6.** Simplify $20\div\left\{5\left[24\div(-6)\right]\right\}$.

We perform operations within the innermost grouping symbols first and work our way outward.

$$-24\div\left\{-3\left[4\div(-2)\right]\right\}=-24\div\left\{-3(-2)\right\}$$
$$=-24\div 6$$
$$=-4$$

| Example | Student Practice |
|---|---|
| **7.** Ions are atoms or groups of atoms with positive or negative electrical charges. An oxide ion has an electrical charge of $-2$, while a magnesium ion has a charge of $+2$. Find the total charge of 8 oxide and 3 magnesium ions.<br><br>We summarize the information. The total charge is equal to the number of oxide ions times the charge of one oxide ion plus the number of magnesium ions times the charge of one magnesium ion.<br><br>$$\text{Total charge} = 8 \times (-2) + 3 \times (+2)$$<br>$$= -16 + 6$$<br>$$= -10$$ | **8.** Use the information from example **7** to answer the following. Find the total charge of 4 oxide ions and 7 magnesium ions. |

**Extra Practice**

**1.** Simplify.

$$2(-5)(7-5)-7$$

**2.** Simplify.

$$\frac{16(-1)-(-4)(-5)}{2\left[-12 \div (-3-3)\right]}$$

**3.** Evaluate $(x+2)^2 - (-y-3)^3$ for $x = -5$ and $y = -2$.

**4.** When Isaac decorated his dining room, he bought a table for $570, 2 armchairs for $125 each, and 4 chairs for $75 each. Write an expression that represents this situation and evaluate it to find out how much Isaac spent.

**Concept Check**

Explain in what order to do the operations to obtain the answer to the problem $3^2 + 5(2-4)$.

**Chapter 2 Integers**
**2.6 Simplifying and Evaluating Algebraic Expressions**

**Vocabulary**
integers   •   commutative   •   distributive property   •   order of operations

1. To multiply $-3(x+1)$, we use the _____.

2. Simplifying algebraic expressions with _____ differs from doing so with whole numbers only in that we must consider the sign of the number when simplifying.

3. To evaluate expressions, we replace the variables with the given numbers and then perform the indicated operations following the _____.

4. Since subtraction is not _____, we must first change all subtractions to additions of the opposite and then rearrange the terms.

| Example | Student Practice |
|---|---|
| **1.** Simplify by combining like terms.<br>$-4x+7y+2x$<br><br>Rearrange terms, then add numerical coefficients of like terms.<br><br>$-4x+7y+2x = -4x+2x+7y$<br>$= (-4+2)x+7y$<br>$= -2x+7y$<br><br>Note: $-2x$ and $7y$ are not like terms. | **2.** Simplify by combining like terms.<br>$-8y+5x+3y$ |
| **3.** Simplify $3x+5y-x$.<br><br>First we change subtraction to addition of the opposite. Then we rearrange terms and add like terms. Note that $-x=-1x$.<br><br>$3x+5y-x = 3x+5y+(-x)$<br>$= 3x+(-1x)+5y$<br>$= 2x+5y$ | **4.** Simplify $5m+2n-m$. |

Vocabulary Answers: 1. distributive property 2. integers 3. order of operations 4. commutative.

| Example | Student Practice |
|---|---|
| **5.** Perform each operation indicated. | **6.** Perform each operation indicated. |

**5.** Perform each operation indicated.

**(a)** $2-3+6$

$$2-3+6 = 2+(-3)+6$$
$$= -1+6$$
$$= 5$$

**(b)** $2x-3x+6x$

$$2x-3x+6x = 2x+(-3x)+6x$$
$$= -1x+6x$$
$$= 5x$$

Notice the similarities with part **(a)**.

**6.** Perform each operation indicated.

**(a)** $-7+3+9$

**(b)** $-7x+3x+9x$

---

**7.** Simplify $3a+8b-9a+3ab-10b$.

First we change subtraction to addition of the opposite.

$$3a+8b+(-9a)+3ab+(-10b)$$

Next we rearrange the terms to group like terms.

$$3a+(-9a)+8b+(-10b)+3ab$$

Then we add coefficients of like terms.

$$[3+(-9)]a+[8+(-10)]b+3ab$$
$$= -6a+(-2b)+3ab$$

Finally, simplify by rewriting addition of the opposite as subtraction.

$$-6a-2b+3ab$$

**8.** Simplify $3m+7n-5mn-6m-9n$.

| Example | Student Practice |
|---|---|
| **9.** Evaluate $\dfrac{\left(x^2 - y\right)}{4}$ for $x = -1$ and $y = -3$. | **10.** Evaluate $\dfrac{\left(m^2 - n\right)}{-5}$ for $m = -3$ and $n = -1$. |

**9.** (continued)

Place parentheses around each variable, then replace $x$ with $-1$ and $y$ with $-3$.

$$\frac{\left[(x)^2 - (y)\right]}{4} = \frac{\left[(-1)^2 - (-3)\right]}{4}$$

Begin by calculating $(-1)^2$. Then, write subtraction as addition of the opposite and simplify.

$$\frac{\left[(-1)^2 - (-3)\right]}{4} = \frac{\left[1 - (-3)\right]}{4}$$
$$= \frac{(1 + 3)}{4}$$
$$= \frac{4}{4}$$
$$= 1$$

---

| **11.** Simplify $-2(y - 4)$. | **12.** Simplify $-7(x - 2)$. |
|---|---|

**11.** (continued)

We distribute the $-2$ over subtraction.

$$-2(y - 4) = -2y - (-2)(4)$$

Next, multiply: $(-2)(4)$.

$$-2y - (-2)(4) = -2y - (-8)$$

We simplify by writing subtraction as addition of the opposite.

$$-2(y - 4) = -2y + 8$$

| Example | Student Practice |
|---|---|

**13.** To find the speed of a free-falling skydiver, we use the formula given below.

The speed of skydiver, $s$, is equal to the initial velocity, $v$, minus the time since start of free fall, $32t$; $s = v - 32t$.

Find the speed of a skydiver at time $t = 5$ seconds if her initial downward velocity $(v)$ is $-7$ feet per second.

We evaluate the formula for the values given: $v = -7$ and $t = 5$.

$$s = v - 32t$$
$$= -7 - 32(5)$$
$$= -7 - 160$$
$$= -167$$

A negative speed means that the object is moving in a downward direction. Therefore, the skydiver is falling 167 feet per second.

**14.** Use the formula in example **13** to answer the following. Find the speed of the skydiver at $t = 6$ seconds if her initial downward velocity $(v)$ is $-2$ feet per second.

**Extra Practice**

**1.** Simplify by combining like terms.
$100 - 3a - 7ab + 5b + 9 + ab$

**2.** Evaluate $12p + 3p^9 - q^2$ for $p = -1$ and $q = 9$.

**3.** Simplify by using the distributive property and combining like terms.
$7(a + 5) + 7$

**4.** Simplify by using the distributive property and combining like terms.
$-6(-2a) + 6(a - 3) + 2(-3a) + 19 + a$

**Concept Check**

State two other ways we can write $-3b + 7$.

# MATH COACH

*Mastering the skills you need to do well on the test.*

Watch the **MATH COACH** videos in MyMathLab® or on You[Tube]™ while you work the problems below. These helpful hints will help you avoid making common errors on test problems.

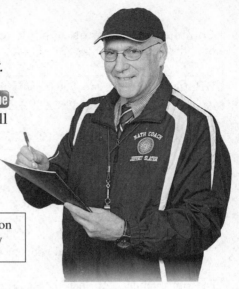

### Adding and Subtracting Integers—Problem 14

$-14-3+(-6)-1$

> **Helpful Hint:** You may find it easier to change all subtraction to addition of the opposite as your first step. This allows you to add numbers in any order. Write out your steps carefully to avoid careless errors.

Look at your work for Problem 14. Examine your steps. As your first step, did you change $-14-3$ to $14+(-3)$?

Yes ____    No ____

As your second step, did you rewrite $(-6)-1$ as a related addition problem? Yes ____    No ____

If you answered No to either question, stop now and complete these steps.

Did you get $-14+(-3)+(-6)+(-1)$ as your next step?

Yes ____    No ____

If you answered No, think about how to rewrite subtraction as addition of the opposite. Recall that the subtraction sign changes to addition, and the number following the subtraction sign is changed to its opposite.

If you answered Problem 14 incorrectly, go back and rework the problem using these suggestions

---

### Multiplying Tow of More Integers—Problem 17    $(-5)(-2)(-1)(3)$

> **Helpful Hint:** As a first step, write down the sign of the product or quotient. Do this *before* you complete any calculations to be sure that you do not forget the sign of the answer.

As a first step, did you identify that there are three negative signs in the problem? Yes ____    No ____

If you answered No, go back and count the number of negative signs again.

Next, did you determine that the product will be negative? Yes ____    No ____

If you answered No, recall that an odd number of negative signs indicates a negative answer, while an even number of negative signs indicates a positive answer.

As your final step, did you multiply the absolute values of the numbers and add the negative sign to your answer? Yes ____    No ____

If you answered No, stop and complete this step again.

Now go back and rework the problem using these suggestions.

## Combining Like Terms with Integer Coefficients—Problem 28

Simplify $-3x + 7xy + 8y - 12x - 11y$.

> **Helpful Hint:** Make sure you *do not* combine two terms with different variables or two terms with a different number of variables. You may only combine like terms. Be careful to avoid sign errors.

Look at your work for Problem 28. Examine your steps.

Did you combine $7xy$ with any other terms?

Yes ＿＿＿   No ＿＿＿

If you answered Yes, stop and look at each term carefully. Notice that the only term with $xy$ as the variable part is $7xy$, so we cannot combine it with any other terms.

Did you combine $-3x$ and $-12x$ and combine $8y$ and $-11y$? Yes ＿＿＿   No ＿＿＿

If you answered No, go back and check your steps. Make sure you did not combine either of the two $x$-terms with either of the $y$-terms.

If you answered Yes, make sure you did not make a *sign error*.

If you answered Problem 28 incorrectly, go back and rework the problem using these suggestions.

---

## Using the Distributive Property with Integers—Problem 29   Simplify $-6(a + 7)$.

> **Helpful Hint:** Be sure to multiply every number or term inside the parentheses by the number outside the parentheses. Then simplify the result. Be careful to avoid sign errors.

Did you multiply both $a$ and 7 by $-6$?

Yes ＿＿＿   No ＿＿＿

If you answered No, then it may help with accuracy to write

arrows as follows: $-6(a + 7)$.

Are both of your terms in the final answer negative?

Yes ＿＿＿   No ＿＿＿

If you answered No, then you made at least one sign error. Try this step again.

Now go back and rework the problem using these suggestions.

## Chapter 3 Introduction to Equations and Algebraic Expressions
## 3.1 Solving Equations of the Form $x + a = c$ and $x - a = c$

**Vocabulary**
opposites • additive inverse property • addition principle of equality • line
ray • angle • line segment • vertex • supplementary angles • adjacent angles

1. A(n) _____ is formed whenever two rays meet at the same endpoint.

2. In geometry a _____ extends indefinitely.

3. A _____ starts at a point and extends indefinitely in one direction.

4. A portion of a line, called a _____, has a beginning and an end.

| Example | Student Practice |
|---|---|
| **1.** Fill in the box with the number that gives the desired result. $$x + 8 + \boxed{\phantom{x}} = x + 0 = x$$ We want the sum of $8 + \boxed{\phantom{x}}$ to equal 0. $$x + 8 + (-8) = x + 0 = x$$ $\qquad \uparrow \qquad\qquad \uparrow$ $$\boxed{8 + (-8) \;=\; 0}$$ Thus, $x + 8 + (-8) = x + 0 = x$. | **2.** Fill in the box with the number that gives the desired result. $$y - 5 + \boxed{\phantom{x}} = y + 0 = y$$ |
| **3.** Solve. $x - 22 = -14$ We want an equation of the form $x = $ some number . Therefore, to get $x$ alone on one side of the equation, add the opposite of $-22$ to both sides of the equation. $$x - 22 + 22 = -14 + 22$$ $$x + 0 = 8$$ $$x = 8$$ | **4.** Solve. $x - 17 = -43$ |

Vocabulary Answers: 1. angle  2. line  3. ray  4. line segment

| Example | Student Practice |
|---|---|
| **5.** Solve and check your solution.<br><br>$2 - 6 = y - 7 + 12$<br><br>First, simplify each side of the equation separately.<br><br>$2 - 6 = y - 7 + 12$<br>$-4 = y + 5$<br><br>Then, add $-5$ to both sides of the equation to get $y$ alone on the right side: some number $= y$.<br><br>$\begin{array}{r} -4 = y + 5 \\ + -5 \quad -5 \\ \hline -9 = y \end{array}$<br><br>Replace $y$ with $-9$ in the original equation and verify that a true statement results. | **6.** Solve and check your solution.<br><br>$7 - 3 = y - 4 + 17$ |
| **7.** Answer parts **(a)** and **(b)**.<br><br>**(a)** Translate into symbols. Angle $y$ measures $40°$ more than angle $x$.<br><br>The phrase "more than" indicates addition, "$+$". Thus, $\angle y = 40° + \angle x$.<br><br>**(b)** Find the measure of $\angle x$ if the measure of $\angle y$ is $95°$.<br><br>First, write an equation and replace $\angle y$ with $95°$. Then, solve the equation.<br><br>$\begin{array}{r} \angle y = 40° + \angle x \\ 95° = 40° + \angle x \\ + -40° \quad -40° \\ \hline 55° = \angle x \end{array}$ | **8.** Answer parts **(a)** and **(b)**.<br><br>**(a)** Translate into symbols. Angle $a$ measures $40°$ less than angle $b$.<br><br>**(b)** Find the measure of $\angle b$ if the measure of $\angle a$ is $60°$. |

| **Example** | **Student Practice** |
|---|---|

**9.** Find $x$ and the measure of $\angle b$ for the pair of supplementary angles in the figure.

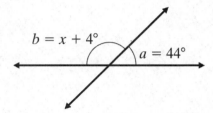

Since $\angle a$ and $\angle b$ are supplementary angles, their sum is 180°. Write an equation as follows.

$\angle a + \angle b = 180°$

Replace $\angle a$ with 44° and $\angle b$ with $x + 4°$.

$44° + (x + 4°) = 180°$

Now, simplify $44° + 4° + x = 48° + x$ and solve the equation.

$$
\begin{array}{rcl}
48° + x & = & 180° \\
+ \quad -48° & & -48° \\
\hline
x & = & 132°
\end{array}
$$

Since $\angle b = x + 4°$, substitute 132° for $x$ to find the measure of $\angle b$.

$\angle b = x + 4°$
$\angle b = 132° + 4° = 136°$

Therefore, $x = 132°$ and $\angle b = 136°$.

**10.** Find $x$ and the measure of $\angle b$ if the measure of $\angle a$ is 56° for the pair of supplementary angles in the figure.

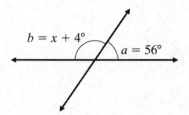

**Extra Practice**

1. Fill in the box with the number that gives the desired result.

$$n + 124 + \boxed{\phantom{0}} = n + 0 = n$$

2. Solve and check your solution.

$$8n - 32 - 7n = 6$$

3. Solve and check your solution.

$$-2 + 4 + x = -3 + 17$$

4. Find the measure of the unknown angle for the given pair of supplementary angles $(\angle a + \angle b = 180°)$.

$$\angle a = ?, \ \angle b = 103°$$

**Concept Check**

To solve the equation $-9 + x = -15$, Damien subtracted 9 from both sides of the equation. Is this correct? Why or why not?

**Chapter 3 Introduction to Equations and Algebraic Expressions**
**3.2 Solving Equations of the Form ax = c**

**Vocabulary**
division principle of equality   •   variables   •   equation   •   multiplication

1.  When we translate a statement into an equation, we must first define the _____ that we are going to use.

2.  The _____ states that if both sides of an equation are divided by the same nonzero number, the results on both sides are equal in value.

3.  Division undoes _____.

4.  Whatever is done to one side of a(n) _____ must be done to the other side.

| Example | Student Practice |
|---|---|
| **1.** Fill in the box with the number that gives the desired result. | **2.** Fill in the box with the number that gives the desired result. |

**Example**

**1.** Fill in the box with the number that gives the desired result.

$$\frac{-5x}{\Box} = 1 \cdot x = x$$

We want the quotient of $\dfrac{-5}{\Box}$ to equal 1.

$$\frac{-5x}{-5} = 1 \cdot x = x$$

$$\uparrow \quad \uparrow$$

$$\boxed{\dfrac{-5}{-5} = 1}$$

**Student Practice**

**2.** Fill in the box with the number that gives the desired result.

$$\frac{-17x}{\Box} = 1 \cdot x = x$$

Vocabulary Answers: 1. variables  2. division principle of equality  3. multiplication  4.

| Example | Student Practice |
|---|---|
| **3.** Solve and check your solution. <br> $7x = -147$ | **4.** Solve and check your solution. <br> $-8x = 152$ |

We want to make $7x = -147$ into a simpler equation, $x = $ some number . The variable, $x$, is multiplied by 7. Dividing both sides of the equation by 7 undoes the multiplication by 7.

$$7x = -147$$

$$\frac{7x}{7} = \frac{-147}{7}$$

$$x = -21$$

Check by replacing $x$ with $-21$ and verifying that a true statement results.

---

**5.** Solve. $3(x \cdot 5) = \dfrac{450}{5}$

**6.** Solve. $6(y \cdot 2) = \dfrac{480}{4}$

First, simplify the equation. Divide: $450 \div 5 = 90$. Then change the order of the factors, regroup the factors, and simplify.

$$3(x \cdot 5) = \frac{450}{5}$$

$$3(x \cdot 5) = 90$$

$$3(5x) = 90$$

$$(3 \cdot 5)x = 90$$

$$15x = 90$$

Finally, dividing both sides by 15 undoes the multiplication by 15.

$$\frac{15x}{15} = \frac{90}{15}$$

$$x = 6$$

| Example | Student Practice |
|---|---|
| **7.** Translate the statement into an equation. There are five times as many dimes $(D)$ as pennies $(P)$ in a coin collection. | **8.** Translate the statement into an equation. There are seven times as many nickels $(N)$ as quarters $(Q)$ in a tip jar. |

**7.** It is helpful to write a sentence that compares the two quantities; "there are more dimes $(D)$ than pennies $(P)$."

Think of a simple comparison of pennies and dimes, such as the case when there is only 1 penny. If there is 1 penny in the collection, then there are 5 dimes.

Now rephrase the statement and translate into an equation; "the number of dimes is five times the number of pennies."

$$D = 5 \cdot P$$

---

**9.** The number of peanuts $(P)$ is triple the number of cashews $(C)$.

    **(a)** Translate the statement into an equation.

    The word triple means three times.
    $P = 3 \cdot C$

    **(b)** Find the number of cashews if there are 27 peanuts.

    Use the equation from part **(a)**. Replace $P$ with 27 and divide both sides by 3.

$$P = 3C$$
$$27 = 3C$$
$$\frac{27}{3} = \frac{3C}{3}$$
$$9 = C$$

**10.** The number of cars $(C)$ is eight times the numbers of bikes $(B)$.

    **(a)** Translate the statement into an equation.

    **(b)** Find the number of bikes if there are 128 cars in the road.

| Example | Student Practice |
|---|---|
| **11.** Lena purchased $x$ shares of stock at $35 per share. She sold all the stock for $56 per share and made a profit of $546. How many shares of stock did Lena purchase? | **12.** Ann purchased $x$ shares of stock at $28 per share. She sold all the stock for $50 per share and made a profit of $374. How many shares of stock did Ann purchase? |

Read the problem carefully and create a Mathematics Blueprint.

$$\boxed{\text{profit}} = \boxed{\text{sale price}} - \boxed{\text{purchase price}}$$
$$546 = \quad 56x \quad - \quad 35x$$

Now combine like terms. Then, divide both sides by 21.

$$546 = 21x$$
$$\frac{546}{21} = \frac{21x}{21}$$
$$26 = x$$

Lena purchased 26 shares of stock. Estimate to see if the answer is reasonable.

**Extra Practice**

**1.** Translate the given statement into an equation.
Miriam is seven times older than her daughter Edie.

**2.** Solve and check your solution.

$$4(5x) = 60$$

**3.** Solve and check your solution.

$$-14 - 28 = 10x - 3x$$

**4.** Matthew moved to an apartment building that's 11 times taller than the house he used to live in. If the apartment building is 253 feet tall, what's the height of the house?

**Concept Check**

Explain in words the steps that are needed to solve $4x + 3(2x) = -20$.

72

**Chapter 3 Introduction to Equations and Algebraic Expressions**
**3.3 Equations and Geometric Formulas**

**Vocabulary**

perimeter • area • parallelogram • parallel lines • height • base • volume

1. The _____ of a rectangular solid is the product of the length times the width times the height.

2. _____ are straight lines that are always the same distance apart.

3. The _____ of a square is the length of one side squared.

4. A _____ is a four-sided figure in which both pairs of opposite sides are parallel.

| **Example** | **Student Practice** |
|---|---|
| **1.** Find the perimeter of a rectangle with $L = 8$ feet and $W = 6$ feet. | **2.** Find the perimeter of a square with sides 14 yards in length. |

We complete the four-step process. Begin by drawing a picture.

$L = 8$ ft
$W = 6$ ft

Now, write the formula.

$$P = 2L + 2W$$

Then, replace $L$ with 8 ft and $W$ with 6 ft.

$$P = 2(8 \text{ ft}) + 2(6 \text{ ft})$$

Finally, simplify to find the perimeter.

$$P = 16 \text{ ft} + 12 \text{ ft}$$
$$= 28 \text{ ft}$$

The perimeter is 28 feet.

Vocabulary Answers: 1. volume  2. parallel lines  3. area  4. parallelogram

| Example | Student Practice |
|---|---|
| **3.** The length of a rectangle is three times the width. If the perimeter of the rectangle is 24 feet, find the width. | **4.** The length of a rectangle is five times the width. If the perimeter of the rectangle is 45 feet, find the width. |

$L = 3W$

$W$

$L = 5W$

$W$

Since we are given a picture, we start by writing the formula, $P = 2L + 2W$. Now, we replace $P$ with 24 and $L$ with the given value $3W$ and solve.

$$24 = 2(3W) + 2W$$

$$24 = 8W$$

$$\frac{24}{8} = \frac{8W}{8}$$

$$3 = W$$

The width of the rectangle is 3 feet.

---

**5.** What is the area of the rug pictured below?

9 yd

3 yd

Think of an array with 3 rows and 9 columns.

9

3

1 yd → This is one square yard.

1 yd

Just as we multiplied the number of rows times the number of columns to find the number of items in an array, we multiply the length times the width to find the area of the rug.

3 yards × 9 yards = 27 square yards

**6.** What is the area of the garden pictured below?

6 ft      9 ft

| Example | Student Practice |
|---|---|
| **7.** Find the area of a rectangle with a length of 3 feet and a width of 2 feet. | **8.** Find the area of a square with a side of 9 feet. |

We complete the four-step process. Begin by drawing a picture.

Now, write the formula, $A = L \cdot W$.

Then, replace $L$ and $W$ with the given values, $A = 3\,\text{ft} \cdot 2\,\text{ft}$.

Finally, simplify. Remember to multiply the units.

$$A = (3 \cdot 2)(\text{ft} \cdot \text{ft})$$
$$= 6\,\text{ft}^2$$

This is read as "six square feet."

| Example | Student Practice |
|---|---|
| **9.** Find the height of a parallelogram with base $= 8$ meters and area area $= 24\ \text{m}^2$. | **10.** Find the base of a parallelogram with area $= 115\ \text{ft}^2$ and height $= 5$ ft. |

Begin by drawing a picture.

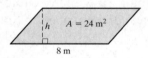

Next, write the formula, $A = bh$.

Then, replace $A$ and $b$ with the values given and solve the equation for $h$.

$$24 = 8h$$
$$\frac{24}{8} = \frac{8h}{8}$$
$$3 = h$$

| Example | Student Practice |
|---|---|
| **11.** Find the unknown side of the rectangular solid. | **12.** Find the unknown side of the rectangular solid. |

$V = 216 \text{ m}^3$

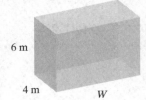

$V = 168 \text{ m}^3$

Since the picture is drawn for us, we first write the formula. Then, replace the variables with the values given, simplify, and solve for $L$.

$$V = L \cdot W \cdot H$$
$$216 = L \cdot 4 \cdot 6$$
$$216 = 24L$$
$$\frac{216}{24} = \frac{24L}{24}$$
$$9 = L$$

The length of the box is 9 m.

**Extra Practice**

**1.** What are the perimeter and area of a square with sides of length 15 yards?

**2.** Find the area of a parallelogram with a height of 15 inches and base of 19 inches.

**3.** Find the volume of a rectangular solid with $L = 8$ feet, $W = 5$ feet, and $H = 2$ feet.

**4.** A plastic feed bin is 4 feet long, 2 feet wide and 2 feet high. How many cubic feet of animal feed can be placed in this bin?

**Concept Check**

Hanna purchased a redwood box at the garden store and filled it with sand.

**(a)** To determine how much sand is needed to fill the box, do you find the area, perimeter, or volume?

**(b)** State the formula you must use.

**(c)** The volume of the box is 200 ft$^3$, the height is 5 in., and the length of the box is double the height. Explain in words how you would find the width of the box.

**Chapter 3 Introduction to Equations and Algebraic Expressions**
**3.4 Performing Operations with Exponents**

**Vocabulary**
product rule for exponents    •    numerical coefficient    •    coefficient
polynomial    •    monomial    •    binomial    •    trinomial

1.  A _____ is an expression that contains terms with variable parts that have only whole number exponents.

2.  A _____ has two terms.

3.  The _____ states that to multiply constants or variables in exponent form that have the same base, add the exponents but keep the base unchanged.

4.  A _____ has three terms.

| **Example** | **Student Practice** |
|---|---|
| **1.** Write $5^4 \cdot 5^3$ as repeated multiplication and then rewrite it with one exponent.<br><br>$$5^4 \quad \cdot \quad 5^3$$<br>$$\overbrace{5 \cdot 5 \cdot 5 \cdot 5} \cdot \overbrace{5 \cdot 5 \cdot 5} = 5^7$$<br><br>Since five appears as a factor 7 times, the exponent is 7, $5^4 \cdot 5^3 = 5^7$. | **2.** Write $7^2 \cdot 7^6$ as repeated multiplication and then rewrite it with one exponent. |
| **3.** Multiply and write the product in exponent form.<br><br>**(a)** $x^3 \cdot x^6$<br><br>$\quad x^3 \cdot x^6 = x^{3+6} = x^9$<br><br>**(b)** $x^7 \cdot x$<br><br>$\quad x^7 \cdot x = x^7 \cdot x^1 = x^{7+1} = x^8$ | **4.** Multiply and write the product in exponent form.<br><br>**(a)** $3^2 \cdot 3^5$<br><br><br>**(b)** $3^2 \cdot 2^5$ |

Vocabulary Answers: 1. polynomial  2. binomial  3. product rule for exponents  4. trinomial

| Example | Student Practice |
|---|---|
| **5.** Multiply. $\left(5x^6\right)\left(7y^4\right)\left(2x^4\right)$ | **6.** Multiply. $\left(7a^2\right)\left(8b^5\right)\left(3a^6\right)$ |

**5.** Multiply. $\left(5x^6\right)\left(7y^4\right)\left(2x^4\right)$

Change the order of factors and regroup them.

$$\left(5x^6\right)\left(7y^4\right)\left(2x^4\right)=(5\cdot 7\cdot 2)\left(x^6\cdot y^4\cdot x^4\right)$$

Multiply the numerical coefficients and variables separately.

$$(5\cdot 7\cdot 2)\left(x^6\cdot y^4\cdot x^4\right)=70x^{10}y^4$$

We cannot use the product rule for exponents to simplify $x^{10}y^4$ because the bases, $x$ and $y$, are not the same.

**6.** Multiply. $\left(7a^2\right)\left(8b^5\right)\left(3a^6\right)$

---

**7.** Multiply. $\left(-7x^2\right)\left(-4y^4\right)$

$$\left(-7x^2\right)\left(-4y^4\right)=(-7)(-4)\left(x^2\cdot y^4\right)$$
$$=28x^2y^4$$

**8.** Multiply. $\left(7a^3\right)\left(-8b^6\right)$

---

**9.** Identify each polynomial as a monomial, binomial, or trinomial.

(a) $2x^2+5$

$2x^2+5$ is a binomial because there are two terms.

(b) $8x$

$8x$ is a monomial because there is one term.

(c) $5x^3+8x-1$

$5x^3+8x-1$ is a trinomial.

**10.** Identify each polynomial as a monomial, binomial, or trinomial.

(a) $9x^2-7x-1$

(b) $2x^5$

(c) $5x^9-10$

| Example | Student Practice |
|---|---|
| **11.** Use the distributive property and then simplify. $3x^2\left(x^5+8x\right)+2x^7$ | **12.** Use the distributive property to simplify. $6x^3\left(x^4-4x\right)+5x^8$ |

Use the distributive property to multiply $3x^2\left(x^5+8x\right)$.

$3x^2 \cdot x^5 + 3x^2 \cdot 8x + 2x^7$

Write $x^5$ as $1x^5$ and $x$ as $x^1$.

$3x^2 \cdot 1x^5 + 3x^2 \cdot 8x^1 + 2x^7$

Multiply numerical coefficients.

$(3\cdot 1)\cdot\left(x^2\cdot x^5\right)+(3\cdot 8)\cdot\left(x^2\cdot x^1\right)+2x^7$

Multiply variables by adding their exponents.

$3x^{2+5}+24x^{2+1}+2x^7$

Simplify and combine the like terms.

$3x^7+24x^3+2x^7=5x^7+24x^3$

Thus, $3x^2\left(x^5+8x\right)+2x^7=5x^7+24x^3$.

| Example | Student Practice |
|---|---|

**13.** Write the area of the rectangle below as an algebraic expression and then simplify.

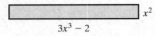

$x^2$

$3x^3 - 2$

First write the formula for area and then replace $L$ and $W$ with the given expressions.

$A = LW$

$\quad = \left(3x^3 - 2\right)\left(x^2\right)$

Then distribute $x^2$ and simplify.

$$\left(3x^3 - 2\right)\left(x^2\right) = \left(3x^3\right)\left(x^2\right) - (2)\left(x^2\right)$$
$$A = 3x^5 - 2x^2$$

**14.** Write the area of the parallelogram below as an algebraic expression and then simplify.

$h = x^4$

$5x^5 - 4x$

**Extra Practice**

**1.** Multiply and write the product in exponent form. $10^3 \cdot 10^5 \cdot 10^9 \cdot 10^0 \cdot 10^1$

**2.** Simplify. $\left(3a^2 - 7a\right)\left(-6a^4\right)$

**3.** Simplify.

$$-4x^3\left(2x + y - 5\right) + 3x^3 y + 4x^4 + y$$

**4.** What is the volume of a box with a length of 20, a width of $3n$, and a height of $n - 1$?

**Concept Check**
Explain in words the steps to simplify each of the following.

**(a)** $2x^4\left(x^2 + y\right)$

**(b)** $\left(-3x^3\right)\left(x^3\right)$

**(c)** Using the results for **(a)** and **(b)** explain the extra steps needed to complete the problem $2x^4\left(x^2 + y\right) + \left(-3x^3\right)\left(x^3\right)$, and then state the answer.

# MATH COACH

*Mastering the skills you need to do well on the test.*

Watch the **MATH COACH** videos in MyMathLab® or on You Tube™ while you work the problems below. These helpful hints will help you avoid making common errors on test problems.

### Solving Equations Using the Addition Principle of Equality— Problem 4

Solve $3x - 2x - 7 = -1 + 6$.

> **Helpful Hint:** When you solve equations, consider the following.
> - Did you simplify each side before you began the process to solve?
> - Did you use the correct principle?
> - Did you remember that whatever you do on one side of the equation, you must do on the other side?
> - Did you take the time to double-check your calculations and $+$ and $-$ signs?

Look at your work for Problem 4. Examine your steps. Did you simplify each side of the equation as your **first step**?
Yes _____ No _____

If you answered No, stop and complete this step.

After simplifying, did you obtain the equation $x - 7 = 5$?
Yes _____ No _____

If you answered No, consider how to identify and combine like terms. Notice that there are like terms to combine on both sides of the equation. Be careful to avoid sign errors.

Did you use the correct principle and add 7 to both sides of the equation?
Yes _____ No _____

If you answered No, go back and complete this step.

If you answered Problem 4 incorrectly, go back and rework the problem using these suggestions.

---

### Solving Equations Using the Division Principle of Equality—Problem 7

Solve $2(4x) = -72$.

> **Helpful Hint:** When you solve equations, consider the following.
> - Did you simplify each side before you began the process to solve?
> - Did you use the correct principle?
> - Did you remember that whatever you do on one side of the equation, you must do on the other side?
> - Did you take the time to double-check your calculations and $+$ and $-$ signs?

Look at your work for Problem 7. Examine your steps. Did you simplify the left side of the equation as your **first step**?
Yes _____ No _____

If you answered No, stop and complete this step.
After simplifying, did you obtain the equation $8x = -72$?
Yes _____ No _____

If you answered No, go back and multiply $2(4x)$ again.

Did you use the correct principle and divide both sides of the equation by 8? Yes _____ No _____

If you answered No, stop and complete this step. Be careful when working with $+$ and $-$ signs.

Now go back and rework the problem using these suggestions.

81

## Multiplying Algebraic Expressions with Exponents—Problem 24

Multiply. Leave your answer in exponent form. $\left(-8x^2\right)\left(-9x^4\right)$

**Helpful Hint:** First multiply the numerical coefficients. Then apply the product rule for exponents by adding the exponents of $x$.

Did you multiply the numerical coefficients $-8$ and $-9$?

Yes _____ No _____

If you answered No, go back and complete this step. Be careful to avoid sign errors.

Did you add the exponents $2 + 4$?

Yes _____ No _____

If you answered No, consider how the product rule for exponents works again. Notice that the two variables being multiplied have the same base but different exponents. Using the product rule, we can add the two exponents.

If you answered Problem 24 incorrectly, go back and rework the problem using these suggestions.

---

## Using the Distributive Property to Multiply a Monomial and a Binomial—Problem 26

Simplify $\left(3x^2 - 5x\right)6x^3$.

**Helpful Hint:** To help with accuracy, you may find it easier to use the commutative property of multiplication to rewrite the problem with the monomial on the left side of the parentheses. Remember to consider the following:

• If a variable does not have an exponent, then the exponent is understood to be 1.

• Multiply both terms inside the parentheses by the monomial.

Look at your work for Problem 26. Examine your steps.

Did you remember to multiply both terms by $6x^3$?
Yes _____ No _____

If you answered No, then stop now and complete these calculations.

Did you multiply $6x^3$ times $3x^2$ to get $18x^5$?
Yes _____ No _____

If you answered No, remember to multiply the numerical coefficients 6 and 3 first and then use the product rule to multiply $x^3$ by $x^2$.

Did you multiply $6x^3$ times $-5x$ to get $-30x^3$?
Yes _____ No _____

If you answered Yes, you forgot that the $x$ in $-5x$ has an exponent of 1. Stop now and apply the product rule again.

Now go back and rework the problem using these suggestions.

Name: _____     Date: _____

Instructor: _____     Section: _____

**Chapter 4 Fractions, Ratio, and Proportion**
**4.1 Factoring Whole Numbers**

**Vocabulary**
divisible • divisible by 2 • divisible by 3 • divisible by 5 • prime number
composite number • factors • prime factors • division ladder • factor tree

1. A _____ is a whole number greater than 1 that is divisible only by itself and 1.

2. A number is _____ if its last digit is 0 or 5.

3. A _____ is a whole number greater than 1 that can be divided by whole numbers other than itself and 1.

4. A number is _____ if the sum of its digits is divisible by 3.

| **Example** | **Student Practice** |
|---|---|
| **1.** Determine if the number is divisible by 2, 3, and/or 5. | **2.** Determine if the number is divisible by 2, 3, and/or 5. |
| **(a)** 234 | **(a)** 225 |
| Divisible by 2 because 234 is even and by 3 since $2+3+4=9$ and 9 is divisible by 3. | |
| **(b)** 38,910 | **(b)** 690 |
| Divisible by 2 because 38,910 is even, by 3 since the sum of the digits is divisible by 3, and by 5 since the last digit is 0. | |
| **3.** State whether each number is prime, composite, or neither. 1, 4, 7, 11, 14, 15, 17, 22, 27, 31, 120 | **4.** State whether each number is prime, composite, or neither. 0, 5, 9, 11, 18, 19, 29, 34, 37, 41, 60 |
| 1 is neither prime nor composite. 4, 14, 15, 22, 27, and 120 are composite. 7, 11, 17, and 31 are prime. | |

Vocabulary Answers: 1. prime number  2. divisible by 5  3. composite number  4. divisible by 3

| Example | Student Practice |
|---|---|
| **5.** Express 20 as a product of prime factors. | **6.** Express 64 as a product of prime factors. |

$20 = 2 \cdot 2 \cdot 5$ or $2^2 \cdot 5$

$20 = 4 \cdot 5$ is not correct because 4 is not a prime number.

---

**7.** Express 28 as a product of prime factors.

**8.** Express 80 as a product of prime factors.

Since 28 is even, it is divisible by the prime number 2. We start the division ladder by dividing 28 by 2.

**Step1** $2\overline{)28}$ with quotient $14$

The quotient 14 is not a prime number. We must continue to divide until the quotient is a prime number.

**Step2** $2\overline{)14}$ with quotient $7$

The quotient 7 is a prime number. Thus all the factors are prime. We are finished dividing. This process is simplified if we write the divisions as follows, placing step 1 on the bottom and moving up the ladder as we divide.

**Step2** $2\overline{)14}$ with quotient $7$
↑ **Step1** $2\overline{)28}$

Now write all the divisors and the quotient as a product of prime factors.

$28 = 2 \cdot 2 \cdot 7$ or $2^2 \cdot 7$

| Example | Student Practice |
|---|---|
| **9.** Express 210 as a product of prime factors. | **10.** Express 336 as a product of prime factors. |

From the divisibility rules we know that 210 is divisible by 2, 3, and 5. We can start with 5.

**Step3**  $2\overline{)14}$ with quotient $7$

**Step2**  $3\overline{)42}$

**Step1**  $5\overline{)210}$

7 is prime, so we are finished dividing.

$$210 = 5 \cdot 3 \cdot 2 \cdot 7 = 2 \cdot 3 \cdot 5 \cdot 7$$

Note that we wrote all factors in ascending order since this is standard notation.

| | |
|---|---|
| **11.** Use a factor tree to express 48 as a product of prime factors. | **12.** Use a factor tree to express 72 as a product of prime factors. |

Write 48 as the product of two factors, $48 = 6 \cdot 8$. Neither 6 nor 8 are prime, so write them as products: $6 = 2 \cdot 3$ and $8 = 2 \cdot 4$. Circle 2 and 3 since they are prime. 4 is not prime, so write it as a product: $4 = 2 \cdot 2$. Circle the factors of 4 since they are prime. Write 48 as a product of the prime factors.

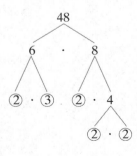

$$48 = 2 \cdot 3 \cdot 2 \cdot 2 \cdot 2 \text{ or } 2^4 \cdot 3.$$

**Extra Practice**

**1.** Determine if the number is divisible by 2, 3, and/or 5.

111,111

**2.** State whether the given number is prime, composite, or neither.

91

**3.** Express the given number as a product of prime factors. Write your answer as powers of prime factors.

98

**4.** Express the given number as a product of prime factors. Write your answer as powers of prime factors.

7200

**Concept Check**

Delroy is having a hard time factoring the number 318.

   **(a)** Explain in words how Delroy can determine one factor of 318.

   **(b)** State this factor.

Name: _____          Date: _____

Instructor: _____          Section: _____

## Chapter 4 Fractions, Ratio, and Proportion
## 4.2 Understanding Fractions

**Vocabulary**

fractions • numerator • denominator • undefined • proper fraction
improper fraction • mixed number

1. A(n) _____ is the sum of a whole number greater than zero and a proper fraction, and is used to describe a quantity greater than 1.

2. A(n) _____ is used to describe a quantity greater than or equal to 1.

3. A(n) _____ is used to describe a quantity less than 1.

4. In mathematics, _____ are a set of numbers used to describe parts of whole quantities.

| Example | Student Practice |
|---|---|
| **1.** Use a fraction to represent the shaded part of each object. | **2.** Use a fraction to represent the shaded part of each object. |

**Example**

**1.** Use a fraction to represent the shaded part of each object.

**(a)**

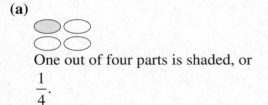

One out of four parts is shaded, or $\dfrac{1}{4}$.

**(b)**

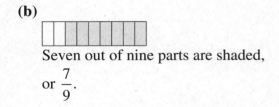

Seven out of nine parts are shaded, or $\dfrac{7}{9}$.

**(c)**

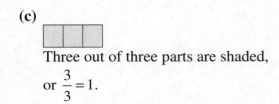

Three out of three parts are shaded, or $\dfrac{3}{3} = 1$.

**Student Practice**

**2.** Use a fraction to represent the shaded part of each object.

**(a)**

○○○○○○

**(b)**

○○○○○○○○○

**(c)**

▭▭▭▭▭▭

Vocabulary Answers: 1. mixed number  2. improper fraction  3. proper fraction  4. fractions

| Example | Student Practice |
|---|---|

**3.** Divide, if possible.

(a) $\dfrac{23}{0}$

Division by 0 is undefined.

(b) $\dfrac{0}{23}$

$\dfrac{0}{23} = 0$

Any fraction with 0 in the numerator and a nonzero denominator equals 0.

**4.** Divide, if possible.

(a) $\dfrac{a}{a},\ a \neq 0$

(b) $\dfrac{0}{19}$

---

**5.** The approximate number of inches of rain that falls during selected periods of one year in Seattle, Washington, is shown by the circle graph. What fractional part of the total yearly rainfall does not occur from July to September?

First find the total rainfall for 1 year.
13 in. + 15 in. + 4 in. + 5 in. = 37 in.

From July to September there were 4 inches of rain out of a total of 37 inches.

Rainfall that does not occur from July to September is 37 in. − 4 in. = 33 in.

Thus, the fractional part of rainfall that does not occur is $\dfrac{33}{37}$.

**6.** Use the circle graph in example **5** to answer the following.

What fractional part of the total yearly rainfall occurs from April to June?

| Example | Student Practice |
|---|---|
| **7.** Identify each as a proper fraction, an improper fraction, or a mixed number. | **8.** Identify each as a proper fraction, an improper fraction, or a mixed number. |

**7.** Identify each as a proper fraction, an improper fraction, or a mixed number.

**(a)** $\dfrac{9}{8}$

Since the numerator is larger than the denominator, it is an improper fraction.

**(b)** $\dfrac{8}{9}$

Since the numerator is less than the denominator, it is a proper fraction.

**(c)** $7\dfrac{3}{4}$

Since a whole number is added to a proper fraction, it is a mixed number.

**8.** Identify each as a proper fraction, an improper fraction, or a mixed number.

**(a)** $\dfrac{17}{6}$

**(b)** $\dfrac{13}{13}$

**(c)** $5\dfrac{6}{7}$

---

**9.** Write $\dfrac{19}{7}$ as a mixed number.

The answer is in the form

$$\text{quotient}\,\dfrac{\text{remainder}}{\text{denominator}}.$$

$$\dfrac{19}{7} \rightarrow 7\overline{)19} \quad \begin{array}{r} 2 \\ \underline{14} \\ 5 \end{array}$$

Here 2 is the quotient and 5 is the remainder. 7 is the denominator from the original fraction.

Thus, $\dfrac{19}{7} = 2\dfrac{5}{7}$.

**10.** Write $\dfrac{29}{4}$ as a mixed number.

| Example | Student Practice |
|---|---|
| **11.** Change $6\frac{1}{2}$ to an improper fraction. | **12.** Change $9\frac{4}{5}$ to an improper fraction. |

To change a mixed number to an improper fraction, multiply the whole number by the denominator of the fraction. Then add this product to the numerator. The result is the numerator of the improper fraction. The denominator does not change.

$$6\frac{1}{2} = \frac{(2\cdot 6)+1}{2} = \frac{12+1}{2} = \frac{13}{2}$$

We can also write the process as follows.

$$6\frac{1}{2} = \frac{2 \text{ times } 6 \text{ plus } 1}{2} = \frac{13}{2}$$

**Extra Practice**

1. A pizza is divided into eight slices and you eat three of them. Write a fraction to describe how much of the pizza you ate.

2. There are 13 girls and 12 boys at the meeting. Write a fraction that describes what part of the meeting are girls.

3. Change the given improper fractions to a mixed number.

$$\frac{83}{8}$$

4. Change the given mixed number to an improper fraction.

$$7\frac{11}{20}$$

**Concept Check**

Explain in words how you change the mixed number $2\frac{3}{4}$ to an improper fraction.

**Chapter 4 Fractions, Ratio, and Proportion**
**4.3 Simplifying Fractional Expressions**

**Vocabulary**
equivalent fractions • reduced to lowest terms • reduce fractions • same value

1. Equivalent fractions look different but have the _____ because they represent the same quantity.

2. To find a(n) _____, we multiply both the numerator and denominator by the same nonzero number.

3. A fraction is considered to be _____ (or written in simplest form) if the numerator and denominator have no common factors other than 1.

4. It is important that you _____ using the method of removing factors of 1.

| Example | Student Practice |
|---|---|
| **1.** Multiply the numerator and denominator of $\frac{3}{4}$ by $3x$ to find an equivalent fraction. | **2.** Multiply the numerator and denominator of $\frac{7}{8}$ by $4x$ to find an equivalent fraction. |
| $$\frac{3}{4} = \frac{3 \cdot 3x}{4 \cdot 3x} = \frac{9x}{12x}$$ | |
| **3.** Write $\frac{3}{4}$ as an equivalent fraction with a denominator of $16x$. | **4.** Write $\frac{5}{7}$ as an equivalent fraction with a denominator of $49x$. |
| $$\frac{3}{4} = \frac{\square}{16x}$$ $$\frac{3 \cdot ?}{4 \cdot ?} = \frac{\square}{16x}$$ 4 times what number equals $16x$?  $4x$ $$\frac{3 \cdot 4x}{4 \cdot 4x} = \frac{12x}{16x}$$ | |

Vocabulary Answers: 1. same value  2. equivalent fraction  3. reduced to lowest terms  4. reduce fractions

| Example | Student Practice |
|---|---|
| **5.** Simplify $\dfrac{-72}{48}$. | **6.** Simplify $\dfrac{81}{-36}$. |

**5.** Simplify $\dfrac{-72}{48}$.

Write the negative sign in front of the fraction.

$$\frac{-72}{48} = -\frac{72}{48}$$

Note that 8 is a common factor of 72 and 48. Write the remaining factors as products of primes and simplify.

$$-\frac{72}{48} = -\frac{\overset{1}{\cancel{8}} \cdot 9}{\underset{1}{\cancel{8}} \cdot 6}$$

$$= -\frac{3 \cdot \overset{1}{\cancel{3}}}{2 \cdot \underset{1}{\cancel{3}}}$$

$$= -\frac{3}{2}$$

**7.** Simplify $\dfrac{150n^2}{200n}$.

Note that 25 is a common factor of the numerator and denominator.

Write all other factors as products of prime numbers to ensure that the fraction is reduced to lowest terms.

$$\frac{150n^2}{200n} = \frac{\overset{1}{\cancel{25}} \cdot 3 \cdot \overset{1}{\cancel{2}} \cdot n \cdot \overset{1}{\cancel{n}}}{\underset{1}{\cancel{25}} \cdot 2 \cdot \underset{1}{\cancel{2}} \cdot 2 \cdot \underset{1}{\cancel{n}}}$$

$$= \frac{3 \cdot n}{2 \cdot 2}$$

$$= \frac{3n}{4}$$

**8.** Simplify $\dfrac{60a^2}{180a}$.

| **Example** | **Student Practice** |
| --- | --- |

**9.** The yearly sales report below shows that the number and type of real estate sales made by Tri-Star Realty.

| Type of Sale | Total Sales |
| --- | --- |
| Condominium | 14 |
| Town home | 21 |
| Single-family home | 45 |

**(a)** What fractional part of the total sales were single-family homes?

First, find the total sales. Then write the fraction and simplify.

Condominium sales + Town home sales + Single-family home sales = Total sales
$14 + 21 + 45 = 80$

$$\frac{\text{single-family home sales}}{\text{total sales}}$$

$$= \frac{48}{80} = \frac{9 \cdot \overset{1}{\cancel{5}}}{16 \cdot \underset{1}{\cancel{5}}} = \frac{9}{16}$$

**(b)** What fractional part of the total sales were not condominiums?

Determine how many sales are not condominiums. Then write the fraction and simplify.

Find the sum: $21 + 45 = 66$.

$$\frac{\text{sales that were not condominiums}}{\text{total sales}}$$

$$= \frac{66}{80} = \frac{33 \cdot \overset{1}{\cancel{2}}}{40 \cdot \underset{1}{\cancel{2}}} = \frac{33}{40}$$

**10.** Use the table in example **9** to answer the following.

**(a)** What fractional part of the total sales were condominiums?

**(b)** What fractional part of the total sales were not town homes?

93

**Extra Practice**

1. Find an equivalent fraction with the given denominator.

$$\frac{5}{8} = \frac{\square}{64a}$$

2. Find an equivalent fraction with the given denominator.

$$\frac{2}{15} = \frac{\square}{45x}$$

3. Simplify $\dfrac{-45xy}{-72y}$.

4. Lane has 24 plants in his house, 18 of which are in his living room. Find a fraction to represent the fractional part of Lane's plants that are in the living room.

**Concept Check**

Can the fraction $\dfrac{105}{231}$ be reduced? Why or why not?

## Chapter 4 Fractions, Ratio, and Proportion
## 4.4 Simplifying Fractional Expressions with Exponents

### Vocabulary
quotient rule  •  raising a power to a power  •  additional power rule  •  product to a power

1. The _____ rule says to keep the same base and multiply the exponents.

2. The _____ states that if a fraction in parentheses is raised to a power, the parentheses indicate that the numerator and denominator are each raised to that power.

3. To raise a(n) _____, raise each factor to that power.

4. The _____ states that if the bases in the numerator and denominator of a fractional expression are the same and $a$ and $b$ are positive integers, then $\dfrac{x^a}{x^b} = x^{a-b}$ if the larger exponent is in the numerator and $x \neq 0$.

| Example | Student Practice |
|---|---|
| **1.** Simplify. Leave your answer in exponent form. | **2.** Simplify. Leave your answer in exponent form. |
| **(a)** $\dfrac{n^9}{n^6}$ | **(a)** $\dfrac{5^{12}}{5^8}$ |
| More factors are in the numerator. | |
| $\dfrac{n^9}{n^6} = n^{9-6} = \dfrac{n^3}{1} = n^3$ | |
| **(b)** $\dfrac{5^8}{5^9}$ | **(b)** $\dfrac{a^6}{a^{11}}$ |
| More factors are in the denominator. | |
| $\dfrac{5^8}{5^9} = \dfrac{1}{5^{9-8}} = \dfrac{1}{5^1}$ or $\dfrac{1}{5}$ | |

Vocabulary Answers: 1. raising a power to a power  2. additional power rule  3. product to a power
4. quotient rule

| Example | Student Practice |
|---|---|

**3.** Simplify $\dfrac{16x^6y^0}{20x^8}$.

Factor 16 and 20.

$$\frac{16x^6y^0}{20x^8} = \frac{\overset{1}{\cancel{4}} \cdot 2 \cdot 2 \cdot x^6 y^0}{\underset{1}{\cancel{4}} \cdot 5 \cdot x^8}$$

Now, $\dfrac{x^6}{x^8} = \dfrac{1}{x^{8-6}}$; $y^0 = 1$.

$$\frac{\overset{1}{\cancel{4}} \cdot 2 \cdot 2 \cdot x^6 y^0}{\underset{1}{\cancel{4}} \cdot 5 \cdot x^8} = \frac{2 \cdot 2 \cdot 1}{5 \cdot x^{8-6}}$$

$$= \frac{4}{5x^2}$$

The leftover $x$ factors are in the denominator.

**4.** Simplify $\dfrac{36a^6b^0}{72a^9}$.

---

**5.** Write $\left(2^4\right)^2$ as a product and then simplify. Leave your answer in exponent form.

$\left(2^4\right)^2$ means $2^4 \cdot 2^4 = 2^{4+4} = 2^8$

**6.** Write $\left(5^3\right)^6$ as a product and then simplify. Leave your answer in exponent form.

| Example | Student Practice |
|---|---|
| **7.** Use the rules for raising a power to a power or a product to a power to simplify. Leave your answer in exponent form. | **8.** Use the rules for raising a power to a power or a product to a power to simplify. Leave your answer in exponent form. |

**(a)** $\left(3^3\right)^3$

Multiply the exponents.

$$\left(3^3\right)^3 = 3^{(3)(3)} = 3^9$$

The base does not change when raising a power to a power.

**(a)** $\left(5^2\right)^4$

**(b)** $\left(x^2\right)^0$

$$\left(x^2\right)^0 = x^{(2)(0)} = x^0 = 1$$

**(b)** $\left(a^0\right)^9$

**(c)** $\left(4x^4\right)^5$

We write 4 as $4^1$.

$$\left(4x^4\right)^5 = \left(4^1 \cdot x^4\right)^5$$

Raise each factor to the power 5.

$$\left(4^1 \cdot x^4\right)^5 = 4^{(1)(5)} \cdot x^{(4)(5)}$$

Finally, multiply exponents.

$$4^{(1)(5)} \cdot x^{(4)(5)} = 4^5 \cdot x^{20}$$

Thus, $\left(4x^4\right)^5 = 4^5 x^{20}$.

**(c)** $\left(5y^3\right)^7$

| Example | Student Practice |
|---|---|
| **9.** Simplify $\left(\dfrac{2}{x}\right)^3$. | **10.** Simplify $\left(\dfrac{x}{5}\right)^4$. |

We must remember to raise both the numerator and the denominator to the power.

$$\left(\frac{2}{x}\right)^3 = \left(\frac{2^1}{x^1}\right)^3 = \frac{2^{(1)(3)}}{x^{(1)(3)}} = \frac{2^3}{x^3} = \frac{8}{x^3}$$

## Extra Practice

**1.** Simplify $\dfrac{10a^3b^7c^0}{15a^5b^5}$. Assume $a$ and $b$ are nonzero. Leave your answer in exponent form.

**2.** Simplify $\dfrac{45x^0y^{17}z^4}{63y^9z^8}$. Assume $y$ and $z$ are nonzero. Leave your answer in exponent form.

**3.** Write $\left(\dfrac{2a^2b^3}{c^4}\right)^2$ as a product and then simplify. Leave your answer in exponent form.

**4.** Write $\left(3y^3z^2\right)^2 \cdot \left(2y^2\right)^3$ as a product and then simplify. Leave your answer in exponent form.

## Concept Check

Explain in words the steps you would need to follow to simplify the expression $\left(\dfrac{6x^2}{3}\right)^3$.

Name: _____    Date: _____

Instructor: _____    Section: _____

**Chapter 4 Fractions, Ratio, and Proportion**
**4.5 Ratios and Rates**

**Vocabulary**

ratio    •    rate    •    unit rate    •    fraction

1.  A(n) _____ is a comparison of two quantities with different units.

2.  Although a ratio can be written in different forms, it is a(n) _____ and therefore should always be simplified (reduced to lowest terms).

3.  A(n) _____ is a comparison of two quantities that have the same units.

4.  When the denominator is 1, we have the rate for a single unit, which is the _____.

| **Example** | **Student Practice** |
|---|---|
| **1.** Write each ratio in simplest form. Express your answer as a fraction. | **2.** Write each ratio in simplest form. Express your answer as a fraction. |
| **(a)** The ratio of 20 dollars to 35 dollars | **(a)** The ratio of 30 feet to 42 feet |
| 20 dollars to 35 dollars $$= \frac{20 \text{ dollars}}{35 \text{ dollars}} = \frac{5 \cdot 4}{5 \cdot 7} = \frac{4}{7}$$ Note that we treat units in the same way we do numbers and variables. | |
| **(b)** $14:21$ $$14:21 = \frac{14}{21} = \frac{7 \cdot 2}{7 \cdot 3} = \frac{2}{3}$$ | **(b)** $32:72$ |

Vocabulary Answers: 1. rate  2. fraction  3. ratio  4. unit rate

| Example | Student Practice |
|---|---|
| **3.** Bertha drove her car 416 miles in 8 hours. Find the unit rate in miles per hour.<br><br>$\dfrac{416 \text{ miles}}{8 \text{ hours}}$ We divide: $8\overline{)416}$ , with $52$ above.<br><br>$\dfrac{416 \text{ miles}}{8 \text{ hours}} = \dfrac{52 \text{ miles}}{1 \text{ hour}}$ or 52 mph | **4.** Lisa travels 95 miles on 5 gallons of gas. Find the unit rate in miles per gallon. |
| **5.** University of Chicago tornado researcher Tetsuya Theodore Fujita cataloged 31,054 tornados in the United States during the 70 years 1916-1985 and found that $\dfrac{7}{10}$ of the tornados occurred in the spring and early summer. Write as a unit rate: the average number of tornados per year that occurred in the month of May. Round your answer to the nearest whole number.<br><br>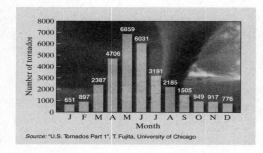Source: "U.S. Tornados Part 1", T. Fujita, University of Chicago<br><br>$\dfrac{6859 \text{ tornados in May}}{70 \text{ years}}$<br><br>Divide to find the unit rate.<br><br>$70\overline{)6859}$ equals $97\frac{69}{70}$ or approximately 98 tornados per year in May | **6.** Use the bar graph in example **5** to answer the following.<br><br>Write as a unit rate: the average number of tornados per year that occurred in the month of July. Round your answer to the nearest whole number. |

| Example | Student Practice |
|---|---|

**7.** Sunshine Preschool has a staffing policy requiring that for every 60 children, there are 3 preschool teachers, and for every 24 children, there are 2 aides.

**(a)** How many children per teacher does the preschool have?

Children per teacher:

$$\frac{\text{children}}{\text{teacher}} \Rightarrow \frac{60 \text{ children}}{3 \text{ teachers}}$$
$$= \frac{20 \text{ children}}{1 \text{ teacher}}$$
or 20 children per teacher

**(b)** How many children per aide does the preschool have?

Children per aide:

$$\frac{\text{children}}{\text{aide}} \Rightarrow \frac{24 \text{ children}}{2 \text{ aides}}$$
$$= \frac{12 \text{ children}}{1 \text{ aide}}$$
or 12 children per aide

**(c)** If there are 60 students at the preschool, how many aides must there be to satisfy the staffing policy?

Since every 12 children require 1 aide, we divide $60 \div 12$ to find how many aides are needed for 60 children.

$60 \div 12 = 5$ aides for 60 children

**8.** A high school has a staffing policy requiring that for every 30 students there are 2 high school teachers, and for every 40 students, there are 2 aides.

**(a)** How many students per teacher does the high school have?

**(b)** How many students per aide does the high school have?

**(c)** If there are 320 students at the high school, how many aides must there be to satisfy the staffing policy?

| Example | Student Practice |
|---|---|

**9.** The Tech Store has black cartridges on sale. A package of 6 sells for $96, and the same brand in a package of 8 sells for $136.

   **(a)** Find each unit price.

$$\frac{\$96}{6} = \$16 \text{ per cartridge;}$$

$$\frac{\$136}{8} = \$17 \text{ per cartridge}$$

   **(b)** Which is the better buy?

   The package of 6 cartridges is the better buy.

**10.** A store has its designer hand towels on sale. A package of 8 sells for $56, and the same brand in a package of 12 sells for $108.

   **(a)** Find each unit price.

   **(b)** Which is the better buy?

**Extra Practice**

**1.** Write the ratio as a fraction in simplest form.

   55 : 10

**2.** Write the ratio as a fraction in simplest form.

   The ratio of $141 to $69

**3.** Cecelia has five dolls and seven stuffed animals on her bed. Write a fraction that describes the ratio of dolls to stuffed animals.

**4.** If five pounds of chicken costs $10.75 and three pounds of pork costs $8.10, which type of meat costs more per pound?

**Concept Check**

A large furniture store determined they needed to have 8 salespeople in the store for every 160 customers. Explain how to determine how many salespeople per customer the store has.

## Chapter 4 Fractions, Ratio, and Proportion
## 4.6 Proportions and Applications

**Vocabulary**

proportion • equality test for fractions • cross product • fractions

1. To determine if a statement is a proportion, we must verify that the _____ in the proportion are equal.

2. By _____ we mean the denominator of one fraction times the numerator of the other fraction.

3. The _____ states that if two fractions are equal, their cross products are equal.

4. A(n) _____ states that two ratios or two rates are equal.

| Example | Student Practice |
|---|---|
| **1.** Translate the statement into a proportion. If 6 pounds of flour cost $2, then 18 pounds will cost $6. <br><br> We can restate as follows: 6 pounds is to $2 as 18 pounds is to $6. <br><br> $$\frac{6 \text{ pounds}}{2 \text{ dollars}} = \frac{18 \text{ pounds}}{6 \text{ dollars}}$$ | **2.** Translate the statement into a proportion. If it takes 5 hours to drive 150 miles, it will take 7 hours to drive 210 miles. |
| **3.** Use the equality test for fractions to see if the fractions are equal. <br><br> $$\frac{2}{11} \overset{?}{=} \frac{18}{99}$$ <br><br> Find the cross products. <br> $\frac{2}{11} \nwarrow \frac{18}{99}$  $99 \cdot 2 = 198$ <br> $\frac{2}{11} \nearrow \frac{18}{99}$  $11 \cdot 18 = 198$ <br> Since $198 = 198$, $\frac{2}{11} = \frac{18}{99}$. | **4.** Use the equality test for fractions to see if the fractions are equal. <br><br> $$\frac{5}{17} \overset{?}{=} \frac{19}{24}$$ |

Vocabulary Answers: 1. fractions  2. cross product  3. equality test for fractions  4. proportion

| Example | Student Practice |
|---|---|
| **5.** Determine if the statement is a proportion. | **6.** Determine if the statement is a proportion. |

**5.**

$$\frac{16 \text{ points}}{35 \text{ games}} \overset{?}{=} \frac{48 \text{ points}}{125 \text{ games}}$$

We check $\frac{16}{35} \overset{?}{=} \frac{48}{125}$ by forming the two cross products.

$$\frac{16}{35} \nwarrow \frac{48}{125} \qquad 125 \cdot 16 = 2000$$

$$\frac{16}{35} \nearrow \frac{48}{125} \qquad 35 \cdot 48 = 1680$$

The two cross products are not equal. Thus, this is not a proportion.

**6.**

$$\frac{18}{45} \overset{?}{=} \frac{54}{135}$$

**7.** Find the value of $n$ in $\frac{n}{24} = \frac{15}{60}$.

First, find the cross products and form an equation.

$$60 \cdot n = 24 \cdot 15$$

Now simplify.

$$60n = 360$$

Finally, divide by 60 on both sides of the equation.

$$\frac{60 \cdot n}{60} = \frac{360}{60}$$

$$n = 6$$

Check by replacing $n$ with 6 in the original equation and verify if a true statement results.

**8.** Find the value of $n$ in $\frac{n}{16} = \frac{32}{64}$.

| Example | Student Practice |
|---|---|
| **9.** Estelle has a fence in her yard around her vegetable garden. The garden is 6 feet wide and 7 feet long. The yard's dimensions are proportional to the garden's. What is the length of the yard if the width is 25 feet? | **10.** Drew's concrete patio in his yard is 8 feet wide and 5 feet long. He wants to enlarge the patio, keeping the dimensions of the new patio proportional to those of the old patio. If he has room to increase the length to 35 feet, how wide should the patio be? |

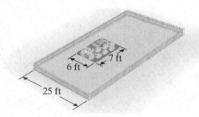

First, set up the proportion. Let the letter $x$ represent the length of the yard.

$$\frac{\text{width of garden}}{\text{length of garden}} = \frac{\text{width of yard}}{\text{length of yard}}$$

$$\frac{6 \text{ ft}}{7 \text{ ft}} = \frac{25 \text{ ft}}{x \text{ ft}}$$

Now solve for $x$.

$$\frac{6}{7} = \frac{25}{x}$$

Find the cross products and then simplify.

$$6x = 7 \times 25$$
$$6x = 175$$
$$\frac{6x}{6} = \frac{175}{6}$$
$$x = 29\frac{1}{6}$$

The length of the yard is $29\frac{1}{6}$ feet.

| Example | Student Practice |
|---|---|
| **11.** Two partners, Cleo and Julie, invest money in their small business at the ratio 3 to 5, with Cleo investing the smaller amount. If Cleo invested $6000, how much did Julie invest? | **12.** Two partners, Mac and Alexander, invest money in their startup business at a ratio of 2 to 5, with Alexander investing the larger amount. If Alexander invested $8000, how much did Mac invest? |

The ratio 3 to 5 represents Cleo's investment to Julie's investment.

$$\frac{3}{5} = \frac{\text{Cleo's investment of \$6000}}{\text{Julie's investment of \$}x}$$

$$\frac{3}{5} = \frac{6000}{x}$$

$$3x = 30,000$$

$$x = 10,000$$

**Extra Practice**

**1.** Use the equality test for fractions to determine if the fractions are equal.

$$\frac{4}{36} \overset{?}{=} \frac{14}{117}$$

**2.** Find the value of $x$ in the given proportion. Check your answer.

$$\frac{4}{13} = \frac{52}{x}$$

**3.** If Tamara takes 15 minutes to stock three shelves of merchandise, how long will it take her to stock 14 shelves?

**4.** Candace reads three books every four weeks. At that rate, how many books will she read in 20 weeks?

**Concept Check**

Justin and Sara share the profits from their business based on the ratio of their investment. The ratio of the investment is 5 to 7, with Sara investing the larger amount. Explain how you would determine how much profit Justin will receive if Sara gets $840.

# MATH COACH

*Mastering the skills you need to do well on the test.*

Watch the **MATH COACH** videos in MyMathLab® or on You Tube™ while you work the problems below. These helpful hints will help you avoid making common errors on test problems.

### Finding the Prime Factors of Whole Numbers—Problem 3

Express as a product of prime factors. 84

**Helpful Hint:** Choose the factoring method (division ladder or factor tree) that you prefer to solve this problem. Make sure all factors in your answer are *prime numbers*. Check your answer by multiplying all factors to be sure this product equals the number you are factoring.

Look at your work for Problem 3. Examine your steps.

*Division ladder:* Did you divide 84 by 4 or 6?
Yes _____ No _____

If you answered Yes, you forgot to use *only prime numbers* as divisors. Stop now and make this correction.

Did you include the final quotient as a factor?
Yes _____ No _____

If you answered No, stop and complete this step.

*Factor Tree:* Are all the factors in your answer prime numbers? Yes _____ No _____

Did you forget to include a prime number listed in the tree as part of your product? Yes _____ No _____

If you answered No to either of these questions, rework the problem and factor each number until all factors are prime. Circle all prime numbers as you factor to be sure you can easily see which factors to include in your answer.

Remember to check your answer by multiplying all the prime factors to be sure that the product is equal to the original number.

If you answered Problem 3 incorrectly, go back and rework the problem using these suggestions.

---

### Finding Equivalent Fractions—Problem 11

Find an equivalent fraction with the given denominator. $\dfrac{4}{9} = \dfrac{?}{27y}$

**Helpful Hint:** Remember to multiply *both* the numerator and denominator by the same nonzero number or expression. The equivalent fraction should have the given denominator of $27y$.

Did you choose 3 as the nonzero number or expression?
Yes _____ No _____

If you answered Yes, your resulting denominator of the equivalent fraction does not equal $27y$. Consider what expression, when multiplied by 9, equals $27y$.

Did you remember to multiply both the numerator and denominator by the same nonzero number or expression?
Yes _____ No _____

If you answered No, stop and make this correction.

Always double-check the product of the original denominator and your nonzero number or expression in the denominator. The result should be equal to the given denominator.

Now go back and rework the problem using these suggestions.

**Use the Quotient Rule for Exponents—Problem 14**   Simplify $\dfrac{y^3 z^4}{y^7 z}$ .

**Helpful Hint:** For this problem, you must apply the quotient rule twice: once for the $y$ variables and once for the $z$ variables. Recall that when an exponent is not written, it is understood to be 1.

Did you notice that the exponent for $y$ is larger in the denominator than in the numerator, and that the exponent for $z$ is larger in the numerator than in the denominator?
Yes _____   No _____

If you answered No, go back and look carefully at the problem again.

Is $y^4$ in the denominator of your answer?
Yes _____   No _____

If you answered No, remember to subtract the exponents of $y$ and to place the resulting $y$ expression in the denominator since the *original exponent* for $y$ is larger in the *denominator*.

Did you remember to subtract exponents for $z$ and obtain the expression $z^3$ ?
Yes _____   No _____

If you answered No, make sure that you write a 1 as the exponent for $z$ in the denominator and complete this step again.

If you answered Problem 14 incorrectly, go back and rework the problem using these suggestions.

---

**Solving Applied Problems Involving Proportions—Problem 28**   A bottle of fertilizer for your lawn states that you need 2 tablespoons to fertilize 400 square feet of lawn. How many tablespoons will you need to fertilize 1600 square feet of lawn?

**Helpful Hint:** When writing and setting up your proportion, remember to write the *unit names* in your proportion to avoid setting up the proportion incorrectly. Make sure that the same unit name appears in the numerator of both fractions and the same unit name appears in the denominator of both fractions.

Did you write $\dfrac{2 \text{ tablespoons}}{400 \text{ square feet}}$ as your first fraction and

$\dfrac{x \text{ tablespoons}}{1600 \text{ square feet}}$ as your second fraction?
Yes _____   No _____

If you answered No, stop and reread the problem to see how these fractions were obtained.

Did you simplify $\dfrac{2}{400}$ before you solved the proportion?
Yes _____   No _____

If you answered No, go back and complete this step again. Simplifying first will make the calculations easier to perform and will help with accuracy.

Remember to include units for $x$ in your final answer.

If you answered Problem 28 incorrectly, go back and rework the problem using these suggestions.

**Chapter 5 Operations on Fractional Expressions**
**5.1 Multiplying and Dividing Fractional Expressions**

**Vocabulary**
common factors  •  reciprocals  •  fractional part  •  inverting the fraction

1. _____ is when the numerator and denominator is interchanged.

2. We multiply in situations that require repeated addition or taking a(n) _____ of something.

3. If the product of two numbers is 1, we say that these two numbers are _____ of each other.

4. Removing a factor of 1 is also referred to as factoring out _____.

| Example | Student Practice |
|---|---|
| **1.** Find $\dfrac{3}{7}$ of $\dfrac{2}{9}$. | **2.** Find $\dfrac{4}{6}$ of $\dfrac{3}{5}$. |

$$\frac{3}{7} \text{ of } \frac{2}{9} = \frac{3}{7} \cdot \frac{2}{9} = \frac{3 \cdot 2}{7 \cdot 9}$$

$$= \frac{3 \cdot 2}{7 \cdot 3 \cdot 3} = \frac{\overset{1}{\cancel{3}} \cdot 2}{7 \cdot \underset{1}{\cancel{3}} \cdot 3} = \frac{2}{21}$$

| | |
|---|---|
| **3.** Multiply $\dfrac{-2}{18} \cdot \dfrac{9}{11}$. | **4.** Multiply $\dfrac{-14}{25} \cdot \dfrac{-15}{28}$. |

When multiplying positive and negative fractions, we determine the sign of the product and then multiply and simplify.

$$\frac{-2}{18} \cdot \frac{9}{11} = (-)$$

$$= -\frac{2 \cdot 9}{18 \cdot 11} = -\frac{2 \cdot 9}{2 \cdot 9 \cdot 11}$$

$$= -\frac{\cancel{2} \cdot \cancel{9}}{\cancel{2} \cdot \cancel{9} \cdot 11} = -\frac{1}{11}$$

Vocabulary Answers: 1. inverting the fraction  2. fractional part  3. reciprocals  4. common factors.

| Example | Student Practice |
|---|---|
| **5.** Multiply $12x^3 \cdot \dfrac{5x^2}{4}$. | **6.** Multiply $\dfrac{2x^4}{5} \cdot \left(10x^3\right)$. |

$$\frac{12x^3}{1} \cdot \frac{5x^2}{4} = \frac{4 \cdot 3 \cdot x^3 \cdot 5 \cdot x^2}{1 \cdot 4}$$

$$= \frac{\cancel{4} \cdot 3 \cdot 5 \cdot x^3 \cdot x^2}{1 \cdot \cancel{4}}$$

$$= \frac{3 \cdot 5 \cdot x^{3+2}}{1} = 15x^5$$

---

**7.** Find the area of a triangle with $b = 12$ in. and $h = 7$ in.

We evaluate the formula with the given values.

$$A = \frac{1}{2}bh$$

$$= \frac{1}{2} \cdot 12 \text{ in.} \cdot 7 \text{ in.}$$

$$= \frac{1 \cdot 12 \text{ in.} \cdot 7 \text{ in.}}{2 \cdot 1 \cdot 1}$$

$$= \frac{1 \cdot 2 \cdot 6 \cdot 7 \text{ in.} \cdot \text{ in.}}{2}$$

$$A = 42 \text{ in.}^2$$

**8.** Find the area of a triangle with $b = 10$ ft and $h = 5$ ft.

---

**9.** Find the reciprocal.

**(a)** $\dfrac{-7}{8}$

To find the reciprocal, we invert the fraction.

$$\frac{-7}{8} \rightarrow \frac{8}{-7} = -\frac{8}{7}$$

**(b)** $6 = \dfrac{6}{1} \rightarrow \dfrac{1}{6} = \dfrac{1}{6}$

**10.** Find the reciprocal.

**(a)** $-\dfrac{1}{3}$

**(b)** $x$

| Example | Student Practice |
|---|---|
| **11.** Divide $\dfrac{-4}{11} \div \left(\dfrac{-3}{5}\right)$. | **12.** Divide $\dfrac{7}{13} \div \left(\dfrac{-15}{17}\right)$. |

There are 2 negative signs in the division. The number 2 is even, so the answer is positive.

$$\frac{-4}{11} \div \left(\frac{-3}{5}\right) = \frac{-4}{11} \cdot \left(\frac{5}{-3}\right)$$
$$= \frac{4 \cdot 5}{11 \cdot 3} = \frac{20}{33}$$

---

**13.** Divide $\dfrac{7x^4}{20} \div \left(\dfrac{-14x^2}{45}\right)$.

**14.** Divide $\dfrac{10x^8}{27} \div \left(\dfrac{-24x^5}{18}\right)$.

$$\frac{7x^4}{20} \div \left(\frac{-14x^2}{45}\right) = \frac{7x^4}{20} \cdot \left(\frac{45}{-14x^2}\right)$$
$$= -\frac{\not{7} \cdot \not{5} \cdot 3 \cdot 3 \cdot x^4}{2 \cdot 2 \cdot \not{5} \cdot 2 \cdot \not{7} \cdot x^2}$$
$$= -\frac{9x^2}{8}$$

---

**15.** Samuel Jensen has $\dfrac{9}{40}$ of his income withheld for taxes and retirement. What amount is withheld each week if he earns \$1440 per week?

**16.** Nancy has a board that is 50 feet long that he wants to cut into 8 equal pieces. How long is each piece?

The key phrase is "$\dfrac{9}{40}$ of his income."
The word "of" often indicates multiplication.

$$\frac{9}{40} \cdot \frac{\$1440}{1} = \$324$$

\$324 is withheld for taxes and retirement each week.

**Extra Practice**

1. Multiply. Be sure your answer is simplified.

$$\frac{-1}{6} \cdot \frac{18}{19}$$

2. Multiply. Be sure your answer is simplified.

$$\left(\frac{-3x}{4}\right) \cdot \left(\frac{6}{7x}\right) \cdot \left(\frac{2}{-5x}\right)$$

3. Divide. Be sure your answer is simplified.

$$\left(\frac{-5}{12}\right) \div \frac{25}{36}$$

4. Divide. Be sure your answer is simplified.

$$27x^4 \div \frac{9}{4x^3}$$

**Concept Check**

Explain how you would divide $\dfrac{-16x^2}{3}$ by $8x$.

Name: _____     Date: _____

Instructor: _____     Section: _____

## Chapter 5 Operations on Fractional Expressions
## 5.2 Multiples and Least Common Multiples of Algebraic Expressions

### Vocabulary

multiples  •  least common multiple  •  common factors  •  build the LCM

1. The smallest of common multiples is called the _____.

2. We should always factor out _____ before we multiply the numerators and denominators; otherwise we must simplify the product.

3. A quicker method is to _____ using the prime factors of each number.

4. To generate a list of _____ of a number, multiply that number by 1, and then by 2, and then by 3, and so on.

| Example | Student Practice |
|---|---|
| **1.** (a) List the first six multiples of $8x$ and the first six multiples of $12x$. $$8x \cdot 1 = 8x, \ 8x \cdot 2 = 16x, \ 8x \cdot 3 = 24x,$$ $$8x \cdot 4 = 32x, \ 8x \cdot 5 = 40x, \ 8x \cdot 6 = 48x$$ $$12x \cdot 1 = 12x, \ 12x \cdot 2 = 24x, \ 12x \cdot 3 = 36x,$$ $$12x \cdot 4 = 48x, \ 12x \cdot 5 = 60x, \ 12x \cdot 6 = 72x$$ (b) Which of these multiples are common to both lists? The multiples common to both lists are $24x$ and $48x$. | **2.** (a) List the first seven multiples of $6y$ and $10y$. (b) Which of these multiples are common to both lists? |
| **3.** Find the LCM of 10 and 15. First, we list some multiples of 10: 10, 20, 30, 40, 50, 60. Next, we list some multiples of 15: 15, 30, 45, 60. We see that both 30 and 60 are common multiples. Since 30 is the smaller of these common multiples, we call 30 the least common multiple (LCM). | **4.** Find the LCM of 6 and 9. |

Vocabulary Answers: 1. least common multiple  2. common factors  3. build the LCM  4. multiples

| Example | Student Practice |
|---|---|
| **5.** Find the LCM of 18, 42, and 45. | **6.** Find the LCM of 14, 24, and 34. |

**5.** Find the LCM of 18, 42, and 45.

Factor each number, then list the requirements for factorization of LCM, then build the LCM.

$18 = 2 \cdot 3 \cdot 3 \rightarrow$ must have a 2 and pair of 3's $\rightarrow \boxed{\text{LCM} = 2 \cdot 3 \cdot 3 \cdot ?}$

$42 = 2 \cdot 3 \cdot 7 \rightarrow$ must have a 2, a 3, and a 7 $\rightarrow \boxed{\text{LCM} = 2 \cdot 3 \cdot 3 \cdot 7 \cdot ?}$

$45 = 3 \cdot 3 \cdot 5 \rightarrow$ must have a pair of 3's and a 5 $\rightarrow \boxed{\text{LCM} = 2 \cdot 3 \cdot 3 \cdot 7 \cdot 5}$

The LCM of 18, 42, and 45 is $2 \cdot 3 \cdot 3 \cdot 7 \cdot 5 = 630$.

---

**7.** Find the LCM of $2x$, $x^2$, and $6x$.

Factor each expression, then list the requirements for factorization of LCM, then build the LCM.

$2x = 2 \cdot x \rightarrow$ must have a 2 and an $x$ $\rightarrow \boxed{\text{LCM} = 2 \cdot x \cdot ?}$

$x^2 = x \cdot x \rightarrow$ must have a pair of $x$'s $\rightarrow \boxed{\text{LCM} = 2 \cdot x \cdot x \cdot ?}$

$6x = 2 \cdot 3 \cdot x \rightarrow$ must have a 2, a 3 and an $x \rightarrow \boxed{\text{LCM} = 2 \cdot x \cdot x \cdot 3}$

The LCM of $2x$, $x^2$, and $6x$ is $2 \cdot x \cdot x \cdot 3 = 6x^2$.

**8.** Find the LCM of $5x$, $20$, and $10x^3$.

| Example | Student Practice |
|---|---|

**9.** Sonia and Leo are tour guides at a castle. Sonia gives a 40-minute tour of the interior of the castle, and Leo gives a 30-minute tour of the castle grounds. There is a 10-minute break after each tour. If tours start at 8 A.M. what is the next time that both tours will start at the same time?

Read the problem carefully and create a Mathematics Blueprint.

We make a chart to help us develop a plan to solve the problem.

| Tours | Number of Minutes after 8 A.M. Tours Start |
|---|---|
| Interior tours start every 50 minutes (40 min+10-min break). | 50, 100, . . . |
| Grounds tours start every 40 minutes (30 min+10-min break). | 40, 80, . . . |

First we factor 40 and 50 into a product of prime factors and then find the LCM.

$40 = 2 \cdot 2 \cdot 2 \cdot 5, \ 50 = 2 \cdot 5 \cdot 5$

$LCM = 2 \cdot 2 \cdot 2 \cdot 5 \cdot 5 = 200$

Both tours will start at the same time 200 minutes after 8 A.M.

Next we change minutes to hours and minutes. Since we need to know how many 60's are in 200, we divide.

200 minutes $\div$ 60 minutes per hour

$= 3$ hours and 20 minutes after 8

$8 + 3$ hours and 20 minutes $= 11{:}20$ A.M.

At 11:20 A.M. both tours will start at the same time.

**10.** Refer to example **9** to complete this problem. Sonia's tour of the interior of the castle is reduced to 35 minutes. Determine the next time that both tours will start at the same time.

**Extra Practice**

**1.** Find the LCM of 20 and 45.

**2.** Find the LCM of 3, 6, and 21.

**3.** Find the LCM $16x$, $70x^2$, and $7x^3$.

**4.** Two security guards patrol a museum each night. Both security guards begin their patrol from the same point. The security guard patrolling the north wing takes 21 minutes to complete his rounds, while the security guard patrolling the south wing takes 18 minutes to complete his rounds. If both security guards leave the same point at 10:00 P.M., at what time will they cross paths?

**Concept Check**

Is $= 3 \times 5 \times 5 \times 7 \times x \times x \times x$ the correct factorization for the LCM of $63x^2$ and $75x^3$? Why or why not?

Name: _____     Date: _____
Instructor: _____     Section: _____

**Chapter 5 Operations on Fractional Expressions**
**5.3 Adding and Subtracting Fractional Expressions**

**Vocabulary**
common denominator • least common denominator • mixed number • equivalent fractions

1. As answers to applications, _____ are generally easier to understand.

2. To add or subtract fractional expressions with different denominators we find the LCD and write _____ that have the LCD as the denominator.

3. When fractions have the same denominator, we say that these fractions have a(n) _____.

4. The _____ of two fractions is the least common multiple of the two denominators.

| Example | Student Practice |
|---|---|
| **1.** Subtract $\dfrac{7}{15} - \dfrac{3}{15}$.  $\dfrac{7}{15} - \dfrac{3}{15} = \dfrac{7-3}{15}$ $= \dfrac{4}{15}$ | **2.** Add $\dfrac{4}{17} + \dfrac{9}{17}$. |
| **3.** Add $\dfrac{-11}{20} + \left(\dfrac{-13}{20}\right)$. $\dfrac{-11}{20} + \left(\dfrac{-13}{20}\right) = \dfrac{-11+(-13)}{20}$ $= \dfrac{-24}{20}$ $= \dfrac{\cancel{4}(-6)}{\cancel{4}(5)}$ $= -\dfrac{6}{5}$ | **4.** Add $\dfrac{-3}{10} + \left(\dfrac{-11}{10}\right)$. |

Vocabulary Answers: 1. mixed numbers  2. equivalent fractions  3. common denominator  4. least common denominator

| Example | Student Practice |
|---|---|
| **5.** Perform the operation indicated.<br><br>**(a)** $\dfrac{6}{y} - \dfrac{2}{y}$<br><br>$\dfrac{6}{y} - \dfrac{2}{y} = \dfrac{6-2}{y} = \dfrac{4}{y}$<br><br>**(b)** $\dfrac{x}{5} + \dfrac{4}{5}$<br><br>$\dfrac{x}{5} + \dfrac{4}{5} = \dfrac{x+4}{5}$ | **6.** Perform the operation indicated.<br><br>**(a)** $\dfrac{3}{5y} + \dfrac{4}{5y}$<br><br><br><br>**(b)** $\dfrac{x}{8} - \dfrac{3}{8}$ |
| **7.** Find the least common denominator (LCD) of the fractions $\dfrac{1}{12}$, $\dfrac{5}{18}$.<br><br>$12 = 2 \cdot 2 \cdot 3,\ 18 = 2 \cdot 3 \cdot 3$<br>$\text{LCD} = 2 \cdot 2 \cdot 3 \cdot 3 = 36$<br><br>The LCD of $\dfrac{1}{12}$ and $\dfrac{5}{18}$ is 36. | **8.** Find the least common denominator of the fractions $\dfrac{1}{20}$, $\dfrac{6}{25}$. |
| **9.** Write the equivalent fraction for $\dfrac{1}{5}$ that has 40 as the denominator. $\dfrac{1}{5} = \dfrac{?}{40}$<br><br>$\dfrac{1}{5} = \dfrac{?}{40}$<br>$\dfrac{1 \cdot 8}{5 \cdot 8} = \dfrac{8}{40}$<br>$\dfrac{1}{5} = \dfrac{8}{40}$ | **10.** Write the equivalent fractions that have 20 as the denominator.<br><br>**(a)** $\dfrac{3}{2} = \dfrac{?}{20}$<br><br><br><br>**(b)** $\dfrac{1}{4} = \dfrac{?}{20}$ |

| Example | Student Practice |
|---|---|

**11.** Perform the operation indicated.

$$\frac{-5}{7} + \frac{3}{4}$$

Find the LCD of $\frac{-5}{7}$ and $\frac{3}{4}$.

$LCD = 28$

Write equivalent fractions.

$$\frac{-5 \cdot 4}{7 \cdot 4} = \boxed{\frac{-20}{28}} \quad \frac{3 \cdot 7}{4 \cdot 7} = \boxed{\frac{21}{28}}$$

Add the numerators of the fractions with common denominators.

$$\frac{-5}{7} + \frac{3}{4} = \boxed{\frac{-20}{28} + \frac{21}{28}} = \frac{1}{28}$$

**12.** Perform the operation indicated.

$$\frac{17}{30} - \frac{7}{18}$$

---

**13.** Add $\frac{7x}{16} + \frac{3x}{32}$.

Find the LCD of $\frac{7x}{16}$ and $\frac{3x}{32}$.

$LCD = 32$

Write equivalent fractions.

$$\frac{7x \cdot 2}{16 \cdot 2} = \boxed{\frac{14x}{32}} \quad \frac{3x}{32} = \boxed{\frac{3x}{32}}$$

Add fractions with common denominators.

$$\frac{7x}{16} + \frac{3x}{32} = \boxed{\frac{14x}{32} + \frac{3x}{32}} = \frac{14x + 3x}{32} = \frac{17x}{32}$$

**14.** Add $\frac{2x}{25} + \frac{x}{5}$.

119

| Example | Student Practice |
|---|---|

**Example**

**15.** Leila finished $\dfrac{1}{8}$ of her English term paper before spring break and $\dfrac{1}{2}$ of the paper during spring break. How much more did she complete during the break than before the break?

The phrase "how much more" indicates that we subtract.

$\dfrac{1}{2} - \dfrac{1}{8}$  The LCD is 8.

$\dfrac{1}{2} = \dfrac{1 \cdot 4}{2 \cdot 4} = \boxed{\dfrac{4}{8}}$  $\dfrac{1}{8} = \boxed{\dfrac{1}{8}}$

Subtract.  $\dfrac{1}{2} - \dfrac{1}{8} = \boxed{\dfrac{4}{8}} - \boxed{\dfrac{1}{8}} = \dfrac{3}{8}$

**Student Practice**

**16.** Frank painted $\dfrac{2}{5}$ of his home red and $\dfrac{2}{9}$ of his home blue. How much more did he paint red than blue.

**Extra Practice**

**1.** Perform the operation indicated. Be sure to simplify your answer. $\dfrac{4}{15} + \left( \dfrac{-2}{15} \right)$

**2.** Perform the operation indicated. Be sure to simplify your answer. $\dfrac{3}{10} + \dfrac{4}{25}$

**3.** Perform the operation indicated. Be sure to simplify your answer. $\dfrac{-3x}{14} - \left( \dfrac{-8x}{21} \right)$

**4.** Tran and Alex were collecting canned food as part of their class food drive. Tran collected $\dfrac{3}{8}$ of the class total and Alex collected $\dfrac{1}{5}$ of the class total. What part of the total number of cans did Tran and Alex collect together?

**Concept Check**

**(a)** What is a common denominator for the fractions $\dfrac{3x}{20}$ and $\dfrac{5x}{6}$?

**(b)** Explain how you would add the fractions.

## Chapter 5 Operations on Fractional Expressions
## 5.4 Operations with Mixed Numbers

### Vocabulary

carrying    •    borrow    •    mixed numbers    •    improper fractions

1. The process used to _____ with mixed numbers is the opposite of carrying with mixed numbers.

2. Change mixed numbers to _____ before multiplying or dividing.

3. To simplify a mixed number with an improper fraction part, use a process similar to _____ with whole numbers.

4. When adding and subtracting _____, we add or subtract the fractions first and then the whole numbers.

| Example | Student Practice |
|---|---|
| 1. Add. $4\frac{1}{8} + 3\frac{3}{8}$ $$4\frac{1}{8}$$ $$+3\frac{3}{8}$$ $$\overline{\quad 7\frac{4}{8} = 7\frac{1}{2}}$$ | 2. Add. $4\frac{1}{7} + 2\frac{5}{7}$ |
| 3. Add. $4\frac{2}{3} + 2\frac{1}{4}$ The LCD of $\frac{2}{3}$ and $\frac{1}{4}$ is 12. $$4\frac{2}{3} \cdot \frac{4}{4} = \; 4\frac{8}{12}$$ $$+2\frac{1}{4} \cdot \frac{3}{3} = +2\frac{3}{12}$$ $$\overline{\qquad\qquad\quad 6\frac{11}{12}}$$ | 4. Add. $7\frac{1}{4} + 5\frac{2}{5}$ |

Vocabulary Answers: 1. borrow  2. improper fractions  3. carrying  4. mixed numbers

| Example | Student Practice |
|---|---|
| **5.** Add. $2\dfrac{5}{7} + 6\dfrac{2}{3}$ | **6.** Add. $5\dfrac{7}{9} + 2\dfrac{3}{4}$ |

The LCD of $\dfrac{5}{7}$ and $\dfrac{2}{3}$ is 21.

$$
\begin{aligned}
2\dfrac{5}{7}\cdot\dfrac{3}{3} = \quad & 2\dfrac{15}{21}\\[4pt]
+6\dfrac{2}{3}\cdot\dfrac{7}{7} = +6\,&\dfrac{14}{21}\\[4pt]
\hline
8\dfrac{29}{21} &= 8 + 1\dfrac{8}{21}\\[4pt]
&= 9\dfrac{8}{21}
\end{aligned}
$$

| Example | Student Practice |
|---|---|
| **7.** Subtract. $7\dfrac{4}{15} - 2\dfrac{7}{15}$ | **8.** Subtract. $6\dfrac{3}{20} - 4\dfrac{7}{20}$ |

We cannot subtract $\dfrac{4}{15} - \dfrac{7}{15}$ without borrowing.

$$
\begin{aligned}
7\dfrac{4}{15} &= 6 + 1\dfrac{4}{15}\\[4pt]
&= 6 + \dfrac{19}{15}
\end{aligned}
$$

$$
\begin{aligned}
7\dfrac{4}{15} = \quad & 6\dfrac{19}{15}\\[4pt]
-2\dfrac{7}{15} = -2&\dfrac{7}{15}\\[4pt]
\hline
& 4\dfrac{12}{15}
\end{aligned}
$$

We simplify: $4\dfrac{12}{15} = 4\dfrac{4}{5}$.

122

| Example | Student Practice |
|---|---|
| **9.** Subtract $8 - 3\frac{1}{4}$. | **10.** Subtract $6 - 2\frac{3}{4}$. |

**9.** Subtract $8 - 3\frac{1}{4}$.

$$8 = \quad 7\frac{4}{4}$$

$$-3\frac{1}{4} = -3\frac{1}{4}$$

$$\overline{\qquad\qquad 4\frac{3}{4}}$$

When we borrowed 1 from 8, we changed the 1 to $\frac{4}{4}$ so the fraction had the same denominator as $\frac{1}{4}$.

**10.** Subtract $6 - 2\frac{3}{4}$.

---

**11.** Multiply $5\frac{5}{12} \cdot 3\frac{11}{15}$.

We change the mixed numbers to improper fractions and then multiply.

$$5\frac{5}{12} \cdot 3\frac{11}{15} = \frac{65}{12} \cdot \frac{56}{15} = \frac{\cancel{5} \cdot 13 \cdot \cancel{4} \cdot 14}{3 \cdot \cancel{4} \cdot \cancel{5} \cdot 3}$$

$$= \frac{182}{9} \text{ or } 20\frac{2}{9}$$

**12.** Multiply $7\frac{4}{5} \cdot 2\frac{6}{7}$.

---

**13.** Divide $2\frac{1}{4} \div (-5)$.

Recall that to divide we invert the second fraction and multiply.

$$2\frac{1}{4} \div (-5) = \frac{9}{4} \div \frac{(-5)}{1}$$

$$= \frac{9}{4} \cdot \left(-\frac{1}{5}\right)$$

$$= -\frac{9}{20}$$

**14.** Divide $4\frac{5}{6} \div (-4)$.

| Example | Student Practice |
|---|---|

**15.** Ester uses a small piece of painted wood as the base for each centerpiece she makes for banquet tables. She has a long piece of word that measures $13\frac{1}{2}$ feet. She needs to cut it into pieces that are $\frac{1}{2}$ foot long for the centerpiece bases. How many centerpiece bases will she be able to cut from the long piece of wood?

**16.** A recipe for pancakes calls for $2\frac{2}{3}$ cups of flour. If Connor only has a $\frac{1}{3}$-cup measuring utensil, how many times must he fill this utensil to get the desired amount of flour?

Draw a picture.

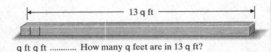

q ft q ft ............ How many q feet are in 13 q ft?

We want to know how many $\frac{1}{2}$s are in $13\frac{1}{2}$, so we must divide $13\frac{1}{2} \div \frac{1}{2}$.

$$13\frac{1}{2} \div \frac{1}{2} = \frac{27}{2} \div \frac{1}{2} = \frac{27}{2} \cdot \frac{2}{1} = 27$$

Ester can make 27 centerpiece bases.

**Extra Practice**

**1.** Add. Simplify the answer. Express as a mixed number. $5\frac{6}{7} + 8\frac{11}{14}$

**2.** Subtract. Simplify the answer. Express as a mixed number. $14\frac{3}{8} - 5\frac{9}{16}$

**3.** Multiply and simplify your answer. $2\frac{3}{5} \cdot 4\frac{5}{8}$

**4.** Divide and simplify your answer. $-2\frac{3}{7} \div \frac{3}{14}$

**Concept Check**

Explain how you would multiply $2\frac{1}{2} \times 3\frac{2}{3}$.

124

**Chapter 5 Operations on Fractional Expressions**
**5.5 Order of Operations and Complex Fractions**

**Vocabulary**
order of operations  •  grouping symbols  •  complex fraction  •  main fraction bar

1. A fraction that contains at least one fraction in the numerator or in the denominator is a(n) _____.

2. We must perform operations above and then below the _____ before we divide.

3. Recall that when we work a problem with more than one operation, we must follow the _____.

4. Although we usually do not write _____ (parentheses or brackets) around the numerator and denominator of a complex fraction, it is understood that they exist.

| **Example** | **Student Practice** |
|---|---|
| **1.** Simplify $\left(\dfrac{2}{3}\right)^2 - \dfrac{2}{9} \cdot \dfrac{1}{3}$. $$\left(\dfrac{2}{3}\right)^2 - \dfrac{2}{9} \cdot \dfrac{1}{3} = \dfrac{4}{9} - \dfrac{2}{9} \cdot \dfrac{1}{3} = \dfrac{4}{9} - \dfrac{2}{27}$$ $$= \dfrac{4 \cdot 3}{9 \cdot 3} - \dfrac{2}{27} = \dfrac{12}{27} - \dfrac{2}{27} = \dfrac{10}{27}$$ | **2.** Simplify $\left(\dfrac{2}{5}\right)^3 + \dfrac{1}{3} \div \dfrac{7}{9}$. |
| **3.** Simplify $\dfrac{(-2)^2 + 8}{\dfrac{2}{3}}$. We must follow the order of operations. $$\dfrac{\left[(-2)^2 + 8\right]}{\left(\dfrac{2}{3}\right)} = \dfrac{12}{\left(\dfrac{2}{3}\right)} = 12 \div \dfrac{2}{3}$$ $$= 12 \cdot \dfrac{3}{2} = \dfrac{\overset{1}{\cancel{2}} \cdot 2 \cdot 3 \cdot 3}{\underset{1}{\cancel{2}}} = 18$$ | **4.** Simplify $\dfrac{\dfrac{3}{4}}{5^2 - (-7)}$. |

Vocabulary Answers: 1. complex fraction  2. main fraction bar  3.order of operations  4. grouping symbols

| Example | Student Practice |
|---|---|

**Example**

**5.** Simplify $\dfrac{\dfrac{x^2}{8}}{\dfrac{x}{4}}$.

Since the main fraction bar indicates division, we can divide the top fraction by the bottom fraction to simplify.

$$\frac{\dfrac{x^2}{8}}{\dfrac{x}{4}} = \frac{x^2}{8} \div \frac{x}{4} = \frac{x^2}{8} \cdot \frac{4}{x}$$

$$= \frac{x^2 \cdot \cancel{4}}{2 \cdot \cancel{4} \cdot x} = \frac{x \cdot x}{2 \cdot x} = \frac{x}{2}$$

**7.** Simplify $\dfrac{\dfrac{2}{3}+\dfrac{1}{6}}{\dfrac{3}{4}-\dfrac{1}{2}}$.

We write parentheses in the numerator and denominator and follow the order of operations.

$$\frac{\left(\dfrac{2}{3}+\dfrac{1}{6}\right)}{\left(\dfrac{3}{4}-\dfrac{1}{2}\right)} = \frac{\left(\dfrac{2\cdot2}{3\cdot2}+\dfrac{1}{6}\right)}{\left(\dfrac{3}{4}-\dfrac{1\cdot2}{2\cdot2}\right)} = \frac{\left(\dfrac{4}{6}+\dfrac{1}{6}\right)}{\left(\dfrac{3}{4}-\dfrac{2}{4}\right)} = \frac{\dfrac{5}{6}}{\dfrac{1}{4}}$$

Now we divide the top fraction by the bottom fraction.

$$\frac{5}{6} \div \frac{1}{4} = \frac{5}{6} \cdot \frac{4}{1} = \frac{5 \cdot \cancel{2} \cdot 2}{3 \cdot \cancel{2}} = \frac{10}{3}$$

**Student Practice**

**6.** Simplify $\dfrac{\dfrac{6}{x}}{\dfrac{2}{x^2}}$.

**8.** Simplify $\dfrac{\dfrac{3}{4}-\dfrac{1}{8}}{\dfrac{7}{9}+\dfrac{5}{6}}$.

| Example | Student Practice |
|---|---|

**9.** A recipe requires $3\frac{2}{3}$ cups of flour to make bread and feed 50 people. How much flour do we need to make bread to feed 120 people?

Since the problem concerns the rate of cups of flour per 50 people, we set up a proportion and solve for the missing number.

$$\frac{3\frac{2}{3}\text{ cups}}{50\text{ people}} = \frac{x\text{ cups}}{120\text{ people}}$$

$$120 \cdot 3\frac{2}{3} = 50x$$

$$120 \cdot \frac{11}{3} = 50x$$

$$\frac{40 \cdot \overset{1}{\cancel{3}} \cdot 11}{\underset{1}{\cancel{3}}} = 50x$$

$$40 \cdot 11 = 50x$$

$$\frac{40 \cdot 11}{50} = \frac{50x}{50}$$

$$\frac{4 \cdot \overset{1}{\cancel{10}} \cdot 11}{5 \cdot \underset{1}{\cancel{10}}} = x$$

$$\frac{44}{5} = x \text{ or } x = 8\frac{4}{5}$$

We need $8\frac{4}{5}$ cups of flour.

**10.** To make 3 bracelets you need $21\frac{3}{4}$ inches of yarn. How much yarn do you need to make 28 bracelets?

127

**Extra Practice**

1. Simplify $\left(\dfrac{3}{5}\right)^3 \cdot \left(\dfrac{1}{3}\right)^3$.

2. Simplify $\left(\dfrac{2}{5} - \dfrac{3}{10}\right)\left(\dfrac{2}{5} + \dfrac{3}{10}\right)$.

3. Simplify $\dfrac{\dfrac{x^2}{3}}{\dfrac{x}{18}}$.

4. Simplify $\dfrac{\dfrac{16}{30} - \dfrac{2}{15}}{\dfrac{14}{40} + \dfrac{9}{20}}$.

**Concept Check**

Explain how you would simplify $\dfrac{1 + 2 \times 3}{\dfrac{1}{2}}$.

**Chapter 5 Operations on Fractional Expressions**
**5.6 Solving Applied Problems Involving Fractions**

**Vocabulary**

fractions     •     improper fractions

1. We must change mixed numbers to _____ before we perform division.

2. For applied problems involving _____, we may need to draw a picture to help us determine which operation to use.

| Example | Student Practice |
|---|---|
| 1. Jason planted a rectangular rose garden in the center of his 26-foot by 20-foot backyard. Around the garden there is a sidewalk that is $3\frac{1}{2}$ feet wide. The garden and sidewalk take up the entire 26-foot by 20-foot yard. What are the dimensions of the rose garden? | 2. Leona wants a rectangular coy pond in the middle of her 32-foot by 28-foot backyard. She wants around the pond to be a rock formation that is $2\frac{1}{4}$ feet wide. The pond and rocks will take up the entire 32-foot by 28-foot yard. |
| Read the problem carefully and create a Mathematics Blueprint. | **(a)** What are the dimensions of the coy pond. |
| We find the length and width of the rose garden. | |
| $L = 26 - \left(3\frac{1}{2} + 3\frac{1}{2}\right)$ | |
| $L = 26 - 7 = 19$ | |
| We find the width of the rose garden. | **(b)** How much will it cost to put a fence around the coy pond if the fencing costs $\$3\frac{1}{2}$ per linear foot? |
| $W = 20 - \left(3\frac{1}{2} + 3\frac{1}{2}\right)$ | |
| $W = 20 - 7 = 13$ | |
| The dimensions of the garden are 19 feet by 13 feet. | |

Vocabulary Answers: 1. improper fractions  2. fractions

| Example | Student Practice |
|---|---|
| **3.** Marian is planning to build a fence on her farm. She determines that she must make 115 wooden fence posts that are each $3\frac{3}{4}$ feet in length. The wood to make the fence posts is sold in 20-foot lengths. How many 20-foot pieces of wood must Marian purchase so that she can make 115 fence posts? | **4.** Martin wishes to make 92 shelves that are each $3\frac{3}{8}$ feet in length. The wood to make the shelves is sold in 10-foot lengths. How many 10-foot pieces of wood must Martin purchase so that he can make 92 shelves? |

**3.** (continued)

Read the problem carefully and create a Mathematics Blueprint.

We must divide to find out how many $3\frac{3}{4}$-foot sections are in 20 feet.

$$20 \div 3\frac{3}{4} = 20 \div \frac{15}{4} = 20 \cdot \frac{4}{15}$$
$$= \frac{4 \cdot \cancel{5} \cdot 4}{\cancel{5} \cdot 3} = \frac{16}{3} \text{ or } 5\frac{1}{3}$$

5 posts can be cut from each 20-foot piece of wood, with some wood left over.

Now we must find how many of the 20-foot pieces are needed. We must find how many groups of 5 are in 115. We divide $115 \div 5$.

$$115 \div 5 = \frac{115}{5} = \frac{\cancel{5} \cdot 23}{\cancel{5}} = 23$$

Marian must purchase 23 pieces of wood.

Check the answer by rounding $3\frac{3}{4}$ to 4 and reworking the problem.

**Extra Practice**

1. Amy and Alex have 55 feet of ribbon. How much ribbon will be left if they use $\frac{1}{5}$ of the ribbon to decorate their raffle table and they use $\frac{5}{7}$ of the ribbon to hang nametags?

2. Allison and Tyler are planning a cookout. They invited 27 people (including themselves) to the cookout. If they estimate that each person can eat $\frac{1}{3}$ pound of meat, $\frac{3}{5}$ pound of potato salad, and $\frac{1}{2}$ pound of fruit, how much meat, salad, and fruit must they order?

3. Plastic tubing is sold in 40-foot bundles. To complete an order, Kieran needs to cut 48 sections of plastic tubing that are each $3\frac{5}{8}$ feet long. How many bundles of plastic tubing will Kieran need to purchase to complete the order?

4. The instructions for making a concrete driveway require $13\frac{1}{3}$ parts cement, 40 parts sand, and 90 parts aggregate. How many parts cement are required to make 8 concrete driveways?

**Concept Check**

Choose the correct operation you must use to answer each of the following questions: Add, Subtract, Multiply, or Divide. You do not need to calculate the answer.

(a) Jason ran $2\frac{1}{3}$ miles, and Lester ran $2\frac{7}{8}$ miles. How much farther did Lester run than Jason?

(b) Beatrice earns $780 per week and has $\frac{1}{13}$ of her paycheck placed in a savings account. How much money does she put in her savings each week?

(c) Samuel has 14 pounds of candy and must place it in $\frac{2}{3}$-pound bags. How many bags can he fill?

## Chapter 5 Operations on Fractional Expressions

## 5.7 Solving Equations of the Form $\dfrac{x}{a} = c$

### Vocabulary

equal        •        multiply        •        reciprocal        •        simplification

1.  To solve an equation of the form $\dfrac{x}{a} = b$, we can _____ both sides of the equation by the same nonzero number.

2.  If both sides of an equation are multiplied by the same nonzero number, the results on both sides are _____ in value.

3.  It is important that we remember to perform any necessary _____ of an equation before we find the solution.

4.  When you multiply a fraction by its _____, the product is 1.

| Example | Student Practice |
|---|---|
| **1.** Solve $\dfrac{x}{-4} = 28$. | **2.** Solve $\dfrac{y}{-3} = -20$. |

Since we are dividing the variable $x$ by $-4$, we can undo the division and get $x$ alone by multiplying by $-4$.

$$\frac{x}{-4} = 28$$

$$\frac{-4 \cdot x}{-4} = 28 \cdot (-4)$$

Simplify: $\dfrac{-4x}{-4} = x$ and $28 \cdot (-4) = -112$.

$$x = -112$$

Be sure that you check your solution.

Vocabulary Answers: 1. multiply  2. equal  3. simplification  4. reciprocal

| Example | Student Practice |
|---|---|

**3.** Solve $\dfrac{x}{2^3} = \dfrac{1}{2} + \dfrac{1}{4}$.

We simplify each side of the equation first and then we find the solution.

$$\frac{x}{2^3} = \frac{1}{2} + \frac{1}{4}$$

Simplify: $2^3 = 8$

$$\frac{x}{8} = \frac{1}{2} + \frac{1}{4}$$

Add: $\dfrac{1}{2} + \dfrac{1}{4} = \dfrac{2}{4} + \dfrac{1}{4} = \dfrac{3}{4}$

$$\frac{x}{8} = \frac{3}{4}$$

We undo the division by multiplying both sides by 8.

$$\frac{8 \cdot x}{8} = \frac{3}{4} \cdot 8$$

Multiply to find the solution: $8 \cdot \dfrac{3}{4} = 6$

$x = 6$

We leave the check for the student.

**4.** Solve $\dfrac{y}{4^2} = \dfrac{1}{4} + \dfrac{1}{8}$.

| Example | Student Practice |
|---|---|
| **5.** Solve for the variable and check your solution. $-\dfrac{3}{4}x = 12$ | **6.** Solve for the variable and check your solution. $-\dfrac{5}{7}y = -10$ |

$$-\dfrac{3}{4}x = 12$$

Multiply both sides of the equation by $-\dfrac{4}{3}$ because $\left(-\dfrac{4}{3}\right)\left(-\dfrac{3}{4}\right) = 1.$

$$\left(-\dfrac{4}{3}\right)\left(-\dfrac{3}{4}\right)x = 12\left(-\dfrac{4}{3}\right)$$

$$1x = -\dfrac{4 \cdot \cancel{3} \cdot 4}{\cancel{3}}$$

$$x = -16$$

Check.

$$-\dfrac{3}{4}x = 12$$

Replace $x$ with $-16$.

$$\left(-\dfrac{3}{4}\right)(-16) \overset{?}{=} 12$$

$$12 = 12$$

**Extra Practice**

1. Solve and check your solution. $\dfrac{x}{8} = 22$

2. Solve and check your solution.

$$\dfrac{x}{-9} = -27 + 12$$

3. Solve and check your solution. $\dfrac{1}{8}x = -14$

4. Solve and check your solution. $\dfrac{5}{7}x = 25$

**Concept Check**

To solve the equation $\dfrac{x}{-5} = 6$, Amy multiplied both sides of the equation by 5 to obtain $x = 30$. Is this correct? Why or why not?

# MATH COACH

*Mastering the skills you need to do well on the test.*

Watch the **MATH COACH** videos in MyMathLab® or on YouTube® while you work the problems below. These helpful hints will help you avoid making common errors on test problems.

**Multiplying Fractions—Problem 6** $\dfrac{8x}{15} \cdot \dfrac{25}{12x^3}$

**Helpful Hint:** First factor out the common factors. Use the correct rule for exponents to simplify the factors $x$ and $x^3$. Then multiply the remaining numerators and multiply the remaining denominators.

Look at your work for Problem 6. Examine your steps.

Did you simplify the fractions before multiplying?
Yes _____  No _____

If you answered No, stop and complete these calculations.

Did you divide 8 and 12 by 2, and 25 and 15 by 5?
Yes _____  No _____

If you answered Yes, make sure you did not stop with 20 in the numerator and 18 in the denominator. The fraction can be simplified further.

Did you use the quotient rule to simplify?
Yes _____  No _____

If you answered No, go back and perform this step.

If you answered Problem 6 incorrectly, go back and rework the problem using these suggestions.

---

**Adding Fractions with Different Denominators—Problem 15** $\dfrac{2}{21} + \dfrac{5}{9}$

**Helpful Hint:** Make sure you understand how to find the LCD. Then rewrite all fractions with the LCD as the denominator. Remember, you do not add the denominators of fractions.

Did you factor 21 into $3 \cdot 7$, and 9 into $3 \cdot 3$?
Yes _____  No _____

If you answered No, stop and complete this step.

Did you get $3 \cdot 3 \cdot 7$ or 63 as the LCD? Yes _____  No _____

If you answered No, review how to find the LCD of 21 and 9.

Did you multiply $\dfrac{2}{21}$ by $\dfrac{3}{3}$ and $\dfrac{5}{9}$ by $\dfrac{7}{7}$ to Rewrite the fractions? Yes _____  No _____

If you answered No, review how to write equivalent fractions using 63 as the LCD. Now go back and rework the problem using these suggestions.

**Following the Order of Operations with Fractions—Problem 19** $\left(\dfrac{2}{3}\right)^2 + \dfrac{1}{2} \cdot \dfrac{1}{4}$

**Helpful Hint:** Write out the rule for the order of operations and refer to it as you complete each step. Then be sure to write down each step. Skipping steps often leads to errors.

Did you square both 2 and 3 in $\left(\dfrac{2}{3}\right)^2$ as your first step?

Yes _____ No _____

Did you multiply $\dfrac{1}{2} \cdot \dfrac{1}{4}$ as your next step?

Yes _____ No _____

If you answered No to either question, go back and make these corrections.

Did you rewrite $\dfrac{4}{9}$ and $\dfrac{1}{8}$ with the LCD 72 as the denominator and then add? Yes _____ No _____

If you answered No, stop and consider why this step must occur before adding the final two fractions.

If you answered Problem 22 incorrectly, go back and rework the problem using these suggestions.

---

**Solving Applied Problems Involving Fractions—Problem 28** Anna wishes to build 2 bookcases, each with 5 shelves. Each shelf is $3\dfrac{1}{2}$ feet long. The wood for the shelves is sold in 10-foot boards. How many boards does Anna need to buy for the shelves?

**Helpful Hint:** Use the Mathematics Blueprint for Problem Solving to help organize your work. Be sure to change mixed numbers to improper fractions before doing any other steps. Then change any division to multiplication by inverting the second fraction.

Did you realize that the problem requires you to perform the calculation 10 divided by $3\dfrac{1}{2}$? Yes _____ No _____

If you answered No, draw a diagram to help you better understand the problem.

Did you change $3\dfrac{1}{2}$ to $\dfrac{7}{2}$ and then rewrite the division as $10 \cdot \dfrac{2}{7}$? Yes _____ No _____

If you answered No to either step, stop and perform these calculations.

Did you think that your calculations were now complete?
Yes _____ No _____

If you answered Yes, go back and read the problem again. Your answer refers to the number of shelves that can be made out of one 10-foot board. You still need to find out how many 10-foot boards are needed to make 10 shelves, each with a length of $3\dfrac{1}{2}$ feet. Make sure that you answer the question asked in the problem.

Now go back and rework the problem using these suggestions.

**Chapter 6 Polynomials**
**6.1 Adding and Subtracting Polynomials**

**Vocabulary**
add polynomials  •  distributive property  •  opposite  •  subtract polynomials

1. To remove parentheses from some expressions such as $(-1)(2x+6)$, use the
   _____ to multiply each term inside the parentheses by $-1$.

2. To _____, combine like terms.

3. To _____, change the sign of each term in the second polynomial and then add.

4. When a negative sign precedes parentheses, we find the _____ of the expression
   by changing the sign of each term inside the parentheses.

| Example | Student Practice |
|---|---|
| **1.** Identify the terms of the polynomial. $xy^2 - 7y^2 - 2x + 5y$ <br><br> We include the sign in front of the term as part of the term. <br><br> Polynomial: $xy^2 - 7y^2 - 2x + 5y$ <br> Terms: $+xy^2, -7y^2, -2x, +5y$ | **2.** Identify the terms of the polynomial. $a^2b + 5b^2 - 6a - 3b$ |
| **3.** Perform the operations indicated. <br><br> $\left(-4x^2 + 5x - 2\right) + \left(3x^2 + 4\right)$ <br><br> We must combine like terms. Rearrange terms so that like terms are grouped together, then add like terms. <br><br> $\left(-4x^2 + 5x - 2\right) + \left(3x^2 + 4\right)$ <br> $= -4x^2 + 3x^2 + 5x - 2 + 4$ <br> $= -1x^2 + 5x + 2$ <br> $= -x^2 + 5x + 2$ | **4.** Perform the operations indicated. <br><br> $\left(5a^2 - 8a + 2\right) + \left(-4a^2 - 3\right)$ |

Vocabulary Answers: 1. distributive property  2. add polynomials  3. subtract polynomials  4. opposite

| Example | Student Practice |
|---|---|
| **5.** Simplify $-(-2a + 5b - 7c)$. | **6.** Simplify $-(3x + 7y - 4z)$. |

Since there is a $-$ in front of the parentheses, we change the sign of each term.

$$-(-2a + 5b - 7c) = 2a - 5b + 7c$$

**7.** Perform the operations indicated.

$$(3x^2 + 5x - 7) - (6x^2 - 8x - 1)$$

A $-$ sign in front of the parentheses indicates we are subtracting. We change the signs of the terms in the second polynomial, and then add.

$$(3x^2 + 5x - 7) - (6x^2 - 8x - 1)$$
$$= 3x^2 + 5x - 7 + (-6x^2) + 8x + 1$$

Simplify by combining like terms:

$$3x^2 - 6x^2 = -3x^2; \ 5x + 8x = +13x;$$
$$-7 + 1 = -6.$$

$$3x^2 + 5x - 7 + (-6x^2) + 8x + 1$$
$$= -3x^2 + 13x - 6$$

**8.** Perform the operations indicated.

$$(x^2 + 5x - 2) - (4x^2 - 5x + 9)$$

| Example | Student Practice |
|---|---|
| **9.** Perform the operations indicated.<br><br>$6x - 3(-4x^2 + 3) - (-2x^2 + x - 5)$ | **10.** Perform the operations indicated.<br><br>$3a - (7a^2 + 2a) - 6(a^2 + 9a - 4)$ |

First we multiply $(-3)$ times the binomial $(-4x^2 + 3)$ to remove parentheses. $-3(-4x^2) = +12x^2$ and $-3(+3) = -9$

$6x - 3(-4x^2 + 3) - (-2x^2 + x - 5)$
$= 6x + 12x^2 - 9 - (-2x^2 + x - 5)$

We remove parentheses and change the sign of each term inside the parentheses.

$6x + 12x^2 - 9 - (-2x^2 + x - 5)$
$= 6x + 12x^2 - 9 + 2x^2 - x + 5$

We combine like terms: $6x - x = 5x$; $12x^2 + 2x^2 = +14x^2$; $-9 + 5 = -4$.

$6x + 12x^2 - 9 + 2x^2 - x + 5$
$= 5x + 14x^2 - 4$

We write the polynomial so that the powers of $x$ decrease as we read from left to right.

$5x + 14x^2 - 4 = 14x^2 + 5x - 4$

**Extra Practice**

1. Identify the terms of the polynomial.

$$5a^4 - 3a^3 + 2a^2 - 7a + 1$$

2. Perform the operations indicated.

$$\left(3y^2 - 2y - 5\right) + \left(4y^2 - 6y + 4\right)$$

3. Simplify $-\left(-4a^4 - 6a^2 + 9\right)$.

4. Perform the operations indicated.

$$7x - 3(x + 4) - (-5x - 9) - (4x + 3)$$

**Concept Check**

Mitchell subtracted two polynomials as follows.

$$\left(-6x^2 + 3x - 1\right) - \left(4x^2 + 2x - 7\right)$$
$$= -6x^2 + 3x - 1 - 4x^2 + 2x - 7$$
$$= -10x^2 + 5x - 8$$

Did Mitchell complete the problem correctly? Why or why not?

**Chapter 6 Polynomials**
**6.2 Multiplying Polynomials**

**Vocabulary**

negative     •     FOIL method     •     distributive property

1. You can use the _____ only when you multiply a binomial times a binomial.

2. When we multiply a binomial times a trinomial, we use the _____ twice since we must multiply each term of the binomial times the trinomial.

3. When multiplying by a _____ monomial, it is a good idea to check the product, verifying that the sign of each term changes.

| Example | Student Practice |
|---|---|
| **1.** Multiply $-4x(2x-6y-7)$. <br><br> We multiply each term by $-4x$. <br><br> $-4x(2x-6y-7)$ <br> $=-4x(2x)-4x(-6y)-4x(-7)$ <br> $=-8x^2+24xy+28x$ | **2.** Multiply $-5a(4a-8b+2)$. |
| **3.** Multiply $(3x^2+x-6)(-7x^3)$. <br><br> We move the monomial to the left side. <br><br> $-7x^3(3x^2+x-6)$ <br><br> We multiply each term by $-7x^3$. <br><br> $-7x^3(3x^2+x-6)$ <br> $=-21x^5-7x^4+42x^3$ <br><br> Since we are multiplying by a negative monomial, check the sign of each term in the product to be sure it changed. | **4.** Multiply $(4z^2-8z+3)(-5z^5)$. |

Vocabulary Answers: 1. FOIL method  2. distributive property  3. negative

| Example | Student Practice |
|---|---|
| **5.** Multiply $(2x+3)(3x^2+5x-1)$. | **6.** Multiply $(4a+7)(2a^2-a+6)$. |

**5.** Multiply $(2x+3)(3x^2+5x-1)$.

We multiply $2x$ times $3x^2+5x-1$ and then $+3$ times $3x^2+5x-1$.

$$(2x+3)(3x^2+5x-1)$$
$$=2x(3x^2+5x-1)+3(3x^2+5x-1)$$
$$=2x\cdot3x^2+2x\cdot5x+2x(-1)$$
$$\quad+3\cdot3x^2+3\cdot5x+3(-1)$$

We multiply.

$$=6x^3+10x^2-2x+9x^2+15x-3$$

We combine like terms.

$$=6x^3+19x^2+13x-3$$

**6.** Multiply $(4a+7)(2a^2-a+6)$.

---

**7.** Use the distributive property to multiply $(x+1)(x+4)$.

We multiply each term of $(x+1)$ times the binomial $(x+4)$. Then we use the distributive property again; multiply $x(x+4)$ and $(+1)(x+4)$.

$$(x+1)(x+4)=x(x+4)+1(x+4)$$
$$=x\cdot x+x\cdot4+1\cdot x+1\cdot4$$
$$=x^2+4x+1x+4$$

Combine like terms.

$$(x+1)(x+4)=x^2+5x+4$$

**8.** Use the distributive property to multiply $(x+5)(x+4)$.

| Example | Student Practice |
|---|---|

**9.** Multiply $(x+4)(x+3)$.

Multiply the First terms, $x$ and $x$, $x \cdot x = x^2$.

Multiply the Outer terms, $x$ and 3, $x \cdot 3 = 3x$.

Multiply the Inner terms, 4 and $x$, $4 \cdot x = 4x$.

Multiply the Last terms, 4 and 3, $4 \cdot 3 = 12$.

Add the results and combine like terms.

$$(x+4)(x+3) = x^2 + 3x + 4x + 12$$
$$= x^2 + 7x + 12$$

**10.** Multiply $(y+5)(y+1)$.

**11.** Multiply $(y-1)(y-7)$.

Pay special attention to the signs of the terms when we multiply.

Multiply the First terms, $y$ and $y$, $y \cdot y = y^2$.

Multiply the Outer terms, $y$ and $-7$, $y \cdot (-7) = -7y$.

Multiply the Inner terms, $-1$ and $y$, $(-1) \cdot y = -1y$.

Multiply the Last terms, $-1$ and $-7$, $(-1)(-7) = +7$.

Add the results and combine like terms.

$$(y-1)(y-7) = y^2 - 8y + 7$$

**12.** Multiply $(x-2)(x-7)$.

| Example | Student Practice |
|---|---|
| **13.** Multiply $(2x+3)(x-1)$. | **14.** Multiply $(x+3)(7x-4)$. |

Be sure to check the sign of each term.

Multiply the First terms, $2x$ and $x$,
$2x \cdot x = 2x^2$.

Multiply the Outer terms, $2x$ and $-1$,
$2x \cdot (-1) = -2x$.

Multiply the Inner terms, $3$ and $x$,
$3 \cdot x = +3x$.

Multiply the Last terms, $3$ and $-1$,
$3 \cdot (-1) = -3$.

Add the results and combine like terms.

$$(2x+3)(x-1) = 2x^2 + x - 3$$

**Extra Practice**

**1.** Use the distributive property to multiply $(2x-1)(x^2 + 3x + 1)$.

**2.** Use FOIL to multiply $(x+7)(x-5)$.

**3.** Use FOIL to multiply $(-x-9)(-2x+9)$.

**4.** Simplify.
$$-2x(x^2 + 3x + 1) + (x-3)(x+4)$$

**Concept Check**

Multiply each of the following.

**1.** $(x+1)(x+2)$ **2.** $(x-1)(x-2)$

**(a)** Explain why the middle terms in each product of 1 and 2 have opposite signs.
**(b)** Explain why the last terms in each product of 1 and 2 have the same sign.

**Chapter 6 Polynomials**
**6.3 Translating from English to Algebra**

**Vocabulary**

expressions     •     equations     •     variable     •     simplify

1. It is sometimes helpful to let the _____ represent the quantity to which things are being compared.

2. Variable _____ can be solved.

3. Variable _____ do not have equal signs.

4. You can _____ an expression; you cannot solve an expression.

| Example | Student Practice |
|---|---|
| **1.** The length of a garden is double the width. Define the variable expressions for the length and width of the garden. | **2.** The width of a pick-up truck is one third the length. Define the variable expressions for the length and width of the pick-up truck. |

Since we are comparing the length to the width, we let the variable represent the width of the garden. We can choose any variable, so we choose $W$.

Let the width $= W$.

Now we write the expression for the length in terms of the width by translating the statement.

Length of garden   is   double   width
        ↓                ↓        ↓         ↓
     length          =      2 ·      W

We define our variable expression as follows.

width $= W$
length $= 2W$

Vocabulary Answers: 1. variable  2. equations  3. expressions  4. simplify

| Example | Student Practice |
|---|---|
| **3.** The second side of a triangle is 2 inches longer than the first; the third side is 8 inches shorter than three times the length of the first side. Define the variable expression for the length of each side of the triangle. | **4.** The second side of a triangle is 5 centimeters longer than the first; the third side is 10 centimeters shorter than twice the length of the first side. Define the variable expression for the length of each side of the triangle. |

Since we are comparing all the sides to the first side, we let the variable represent the length of the first side.

Let the length of the first side $= f$.

second side   is   2 longer than first side
    ↓         ↓              ↓

second side   =       $f + 2$

To find the expression for the third side, we can separate the sentence "The third side is 8 inches shorter than 3 times the first side" into two phrases and translate each phrase separately.

third side   is   8 inches shorter than
    ↓         ↓              ↓

third side   =       $\square - 8$

3   times   the first
↓     ↓        ↓

3   ·      $f$    $-8$

third side $= 3f - 8$

We define our variable expressions as follows.

length of the first side $= f$

length of the second side $= (f + 2)$

length of the third side $= (3f - 8)$

| Example | Student Practice |
|---|---|

**5.** A company's profit for the first quarter is three times the profit for the second quarter. The profit for the third quarter is $22,100 more than the profit for the second quarter.

   **(a)** Write the variable expression for the profit for each quarter.

      We are comparing the first- and third-quarter profits to the second-quarter profit, so we let the variable represent the second-quarter profit.

      second-quarter profit $= x$

      Now we write the first-quarter profit in terms of the second.

      The profit for the first quarter is three times the profit for the second quarter.

      first-quarter profit $= 3 \cdot x$

      Now we write the third-quarter profit in terms of the second quarter.

      The profit for the third quarter is $22,100 more than the profit for the second quarter.

      third-quarter profit $= (x + 22{,}100)$

   **(b)** Write the following phrase using math symbols: The profit for the first quarter minus the profit for the third quarter plus the profit for the second quarter. Simplify.

      $3x - (x + 22{,}100) + x = 3x - 22{,}100$

**6.** A company's profit for the first quarter is four times the profit for the third quarter. The profit for the second quarter is $25,000 more than the profit for the third quarter.

   **(a)** Write the variable expression for the profit for each quarter.

   **(b)** Write the following phrase using math symbols: The profit for the third quarter minus the profit for the second quarter plus the profit for the first quarter. Simplify.

**Extra Practice**

1. The price of a new vehicle is $10,100 more than the price of a used vehicle of the same model. Define the variable expressions for the cost of each vehicle.

2. The width of a rectangle is 15 inches shorter than twice the length. Define the variable expressions for the length and width of the rectangle.

3. The second side of a triangle is 4 inches longer than the first. The third side is 6 inches shorter than three times the first. Define the variable expressions for the length of each side of the triangle.

4. Brandon has 350 more hockey cards in his collection than Sheldon. Trevor has 180 fewer hockey cards in his collection than Sheldon.
   (a) Define the variable expressions for the numbers of hockey cards in Brandon's, Sheldon's, and Trevor's collections.
   (b) Write the following phrase using math symbols: The number of hockey cards in Brandon's collection plus the number of hockey cards in Trevor's collection minus the number of hockey cards in Sheldon's collection.
   (c) Simplify the expression from part (b).

**Concept Check**

The width of a box is triple the height. The length of the box is six inches shorter than twice the height.
(a) Which side of the box will you choose to let the variable represent: length, width, or height? Explain why.
(b) Define the variable expressions for each side of the box.

Name: _____     Date: _____
Instructor: _____     Section: _____

**Chapter 6 Polynomials**
**6.4 Factoring Using the Greatest Common Factor**

**Vocabulary**

common factors   •   greatest common factor   •   divide   •   factor an expression

1. To factor out the GCF we can multiply the expression by the GCF and then _____ the expression by the GCF.

2. We call 1, 2, 3, and 6 _____ of 6 and 12.

3. To _____, we find an equivalent expression written as a product.

4. We call the largest common factor the _____.

| Example | Student Practice |
|---|---|
| **1.** Find the GCF of 8 and 16. <br><br> Think of a factor, the largest factor, that will divide into both 8 and 16. The largest common factor is 8, therefore the GCF of 8 and 16 is 8. | **2.** Find the GCF of 10 and 5. |
| **3.** Find the GCF of 8, 20, and 28. <br><br> First we write each number as a product of prime factors in exponent form. Then we identify the common factors and use the smallest power that appears on these factors to find our GCF. <br><br> $8 = 2 \cdot 2 \cdot 2 \qquad = 2^3$ <br> $20 = 2 \cdot 2 \cdot \quad 5 \quad = 2^2 \cdot 5$ <br> $28 = 2 \cdot 2 \cdot \qquad 7 = 2^2 \cdot 7$ <br><br> Notice that $2 \cdot 2$ or $2^2$ is common to 8, 20, 28. <br><br> The greatest common factor of 8, 20, 28 is 4. | **4.** Find the GCF of 15, 20, and 30. |

Vocabulary Answers: 1. Divide  2. common factors  3. factor an expression  4. greatest common factor

| Example | Student Practice |
|---|---|
| **5.** Find the GCF. | **6.** Find the GCF. |
| **(a)** $15x^2 y + 18x^3$ | **(a)** $6x^5 y^2 + 10x^2$ |
| Rewrite each term using prime factors written in exponent form, then identify the common prime factors and use the smallest power that appears on each of these factors. | |
| $15x^2 y = 3^1 \cdot 5 \cdot x^2 \cdot y$ | |
| $18x^3 = 3^2 \cdot 2 \cdot x^3$ | |
| The GCF of $15x^2 y + 18x^3$ is $3x^2$. | |
| **(b)** $a^4 bc + a^2 b^2$ | **(b)** $x^3 y^2 z + y^7 z^4$ |
| $a^4 bc = a^4 \cdot b^1 \cdot c$ | |
| $a^2 b^2 = a^2 \cdot b^2$ | |
| The GCF of $a^4 bc + a^2 b^2$ is $a^2 b$. | |
| **7.** Factor $12x + 15$ and check your solution. | **8.** Factor $20x + 35$ and check your solution. |
| 3 is the greatest common factor of $12x + 15$. | |
| Write each term as a product and factor out the GCF, 3. | |
| $12x + 15 = 3 \cdot 4x + 3 \cdot 5$ | |
| $\phantom{12x + 15} = 3(4x + 5)$ | |
| Check. We multiply using the distributive property. | |
| $3(4x + 5) = 3 \cdot 4x + 3 \cdot 5$ | |
| $\phantom{3(4x + 5)} = 12x + 15$ | |

| Example | Student Practice |
|---|---|
| **9.** Factor $8x - 12y + 16$ and check your solution. | **10.** Factor $12x + 24y - 30$ and check your solution. |

**9.**

4 is the greatest common factor of $8x - 12y + 16$.

Write each term as a product and factor out the GCF, 4.

$$8x - 12y + 16 = 4 \cdot 2x - 4 \cdot 3y + 4 \cdot 4$$
$$= 4(2x - 3y + 4)$$

We should always check our answer using the distributive property.

$$4(2x - 3y + 4) = 8x - 12y + 16$$

---

**11.** Factor $5xy^2 + 15xy^3$.

We find the GCF of the expression.

$$5xy^2 = \phantom{3\cdot} 5 \cdot x \cdot y^2$$
$$15xy^3 = 3 \cdot 5 \cdot x \cdot y^3$$
$$\downarrow \downarrow \downarrow$$
$$\text{The GCF} = \phantom{3\cdot} 5 \cdot x \cdot y^2 = 5xy^2$$

We factor out the GCF from the expression.

$$\text{GCF}\left(\frac{5xy^2}{\text{GCF}} + \frac{15xy^3}{\text{GCF}}\right)$$

$$5xy^2 + 15xy^3 = 5xy^2\left(\frac{5xy^2}{5xy^2} + \frac{15xy^3}{5xy^2}\right)$$
$$= 5xy^2(1 + 3y)$$

The check is left to the student.

**12.** Factor $14x^2y + 35x^4y^3$.

## Extra Practice

**1.** Find the GCF for the expression.

$x^4 y^5 - x^3 y^6$

**2.** Factor $3x - 9$. Check by multiplying.

**3.** Factor $30x - 12y + 18$. Check by multiplying.

**4.** Factor $27x^3 y^2 - 34a^3 b^3$. Check by multiplying.

## Concept Check

For the expression $12xy + 16x$

**(a)** Is $xy$ part of the GCF? Why or why not?

**(b)** State the GCF.

**(c)** Factor $12xy + 16x$.

# MATH COACH

*Mastering the skills you need to do well on the test.*

Watch the MATH COACH videos in MyMathLab® or on You Tube™
while you work the problems below. These helpful hints will
help you avoid making common errors on test problems.

## Subtracting Polynomials—Problem 5

Perform the operations indicated. $(-7p-2)-(3p+4)$

> **Helpful Hint:** Be careful when subtracting expressions. Students often
> forget to change the sign of *each term* in the second polynomial. Take
> extra time and check your work to be sure you did not make this error.

After removing parentheses, did you get $-3p+4$ as the last
two terms? Yes _____ No _____

If you answered Yes, you forgot to change the sign of every
term in the second polynomial. Stop now and make this
correction.

Did you get $-7p-3p=-10p$ and $-2-4=-6$ after
removing parentheses? Yes _____ No _____

If you answered No, consider grouping like terms together
before subtracting. Then go back and complete this step
again.

If you answered any of Problem 5 incorrectly,
go back and rework the problem using these
suggestions.

---

## Multiplying Binomials Using FOIL—Problem 13  Multiply $(x+3)(x-2)$.

> **Helpful Hint:** To help with accuracy, try to draw arrows when following the FOIL method. Pay particular
> attention to the sign of each number to avoid errors.

Using FOIL, did you get $\mathbf{F}=x^2$, $\mathbf{O}=-2x$, $\mathbf{I}=+3x$, and
$\mathbf{L}=-6$? Yes _____ No _____

If you answered No, check to see if you made a sign error or
used the FOIL method incorrectly.

Is the middle term of your answer equal to $+x$?
Yes _____ No _____

If you answered No, look at your work for adding the inner
and outer terms. It may help with accuracy to write out the
step $-2x+3x=x$.

Now go back and rework the problem using
these suggestions.

**Writing Variable Expressions When Comparing Two or More Quantities—Problem 17** The width of a piece of wood is 3 inches shorter than the length. Define the variable expressions for the length and width of the piece of wood.

> **Helpful Hint:** It is a good idea to let the variable represent the quantity to which things are being compared. When writing expressions, remember that math symbols are not always written in the same order as they are read in the statement.

Did you let the variable represent the length?

Yes _____ No _____

If you answered No, stop now and make this correction.

Did you write the variable expression: width $= 3 - L$ ?
Yes _____ No _____

If you answered Yes, go back and reread the problem carefully. Notice that the phrase "is shorter than the length" means 3 taken away from $L$.

Remember that the question asks for the variable expressions for both length and width, so you will need to write both of these answers.

If you answered any of Problem 17 incorrectly, go back and rework the problem using these suggestions.

---

**Factoring Out the GCF from the Polynomial—Problem 25**    Factor $2x^2 y - 6xy^2$ .

> **Helpful Hint:**
> - Write each number as a product using *only* prime factors. Then align all common prime factors.
> - When forming the GCF, choose factors that are common to every term and use the smallest power on these common factors.
> - After factoring, double-check to see if the product is factored completely.

Did you factor each term as follows?

$$2x^2 y = 2^1 \cdot \quad x^2 \cdot y^1$$
$$6xy^2 = 2^1 \cdot 3^1 \cdot x^1 \cdot y^2$$

*Notice that common factors are aligned.* ↑    ↑ ↑

Yes _____ No _____

If you answered No, go back and perform this step again.

Did you factor $2x^2 y - 6xy^2$ into $2\left(x^2 y - 3xy^2\right)$ ?

Yes _____ No _____

If you answered Yes, the expression is *not factored completely*. To avoid this situation, you must find the correct GCF. Align common factors as shown in the first step. Then you will see that 2, $x$, and $y$ are common to both terms and therefore part of the GCF.

Did you choose $2x^2 y^2$ as the GCF?
Yes _____ No _____

If you answered Yes, check your work again. You *did not use* the *smallest* power on each common factor: $2^1$, $x^1$, and $y^1$.

Now go back and rework the problem using these suggestions.

**Chapter 7 Solving Equations**
**7.1 Solving Equations Using One Principle of Equality**

**Vocabulary**

addition principle of equality • multiplication principle of equality
division principle of equality • variable

1. The goal when solving an equation is to get the _____ alone on one side of the equation.

2. The _____ states that if both sides of the equation are divided by the same nonzero number, the results on each side are equal in value.

3. The _____ states that if both sides of an equation are multiplied by the same nonzero number, the results on each side are equal in value.

4. The _____ states that if the same number or variable term is added to both sides of the equation, the results on each side are equal in value.

| Example | Student Practice |
|---|---|
| **1.** Solve $-12-15=9x+15-8x-6$. | **2.** Solve $-13-19=7y+14-6y-1$. |

First simplify each side of the equation by combining like terms. Simplify the left side first and then the right.

$$-12-15=9x+15-8x-6$$
$$-27=9x+15-8x-6$$
$$-27=x+9$$

Now solve for $x$. The opposite of $+9$ is $-9$. Add $-9$ to both sides of the equation.

$$-27=x+9$$
$$\underline{+\quad -9\qquad -9}$$
$$-36=x$$

Vocabulary Answers: 1. variable  2. division principle of equality  3. multiplication principle of equality
4. addition principle of equality

| Example | Student Practice |
|---|---|
| **3.** Solve $\dfrac{y}{-5} = -10 + 2^3$ and check your solution. | **4.** Solve $\dfrac{a}{-3} = 2^4 - 12$ and check your solution. |

First simplify each side of the equation by evaluating the exponent and then combining the constants.

$$\frac{y}{-5} = -10 + 2^3$$

$$\frac{y}{-5} = -10 + 8$$

$$\frac{y}{-5} = -2$$

Since $y$ is divided by $-5$, we can undo the division and obtain $y$ alone by multiplying both sides of the equation by $-5$.

$$\frac{-5 \cdot y}{-5} = -2 \cdot (-5)$$

$$y = 10$$

Check the solution by replacing $y$ with 10 and verifying that a true statement results.

$$\frac{y}{-5} = -10 + 2^3$$

$$\frac{10}{-5} \overset{?}{=} -10 + 2^3$$

$$-2 = -2$$

Since the resulting statement is true, the solution is $y = 10$.

| Example | Student Practice |
|---|---|
| **5.** Solve $\dfrac{-20}{5} = -2(5x) + 7x$ and check your solution. | **6.** Solve $\dfrac{18}{-9} = 6x + 5(-4x)$ and check your solution. |

We begin by simplifying each side of the equation.

$$\frac{-20}{5} = -2(5x) + 7x$$
$$-4 = -2(5x) + 7x$$
$$-4 = -10x + 7x$$
$$-4 = -3x$$

Now that both sides of the equation are simplified, we use the division principle of equality to transform the equation into the form *some number* $= x$.

Dividing by $-3$ on both sides undoes the multiplication by $-3$.

$$\frac{-4}{-3} = \frac{-3x}{-3}$$
$$\frac{4}{3} = x$$

Check the solution by replacing $x$ with $\dfrac{4}{3}$ and verifying that a true statement results.

$$\frac{-20}{5} \overset{?}{=} -2\left(5 \cdot \frac{4}{3}\right) + 7 \cdot \frac{4}{3}$$
$$-4 \overset{?}{=} -2\left(\frac{20}{3}\right) + \frac{28}{3}$$
$$-4 \overset{?}{=} -\frac{40}{3} + \frac{28}{3}$$
$$-4 = -4$$

| Example | Student Practice |
|---|---|
| **7.** Solve $-x = 36$. | **8.** Solve $-x = 49$. |

We can rewrite $-x$ as $-1x$. Then we can use the division principle of equality to solve for x.

$$-1x = 36$$
$$\frac{-1x}{-1} = \frac{36}{-1}$$
$$x = -36$$

### Extra Practice

**1.** Solve and check your solution.

$$-12 - 3 = 6x - 10 - 5x + 3$$

**2.** Solve and check your solution.

$$\frac{a}{-4} = 12 - 4^2$$

**3.** Solve and check your solution.

$$\frac{2}{3}x = 2^3 - 6$$

**4.** Solve and check your solution.

$$\frac{10}{-2} = 4(-3x) + 6x$$

### Concept Check

Which of the following equations would you solve by dividing by 7 on both sides of the equation? Explain why this operation is used to solve for $x$.

**(a)** $x - 7 = -21$

**(b)** $\frac{x}{7} = -21$

**(c)** $7x = -21$

**(d)** $x + 7 = -21$

Name: _____     Date: _____

Instructor: _____     Section: _____

## Chapter 7 Solving Equations
## 7.2 Solving Equations Using More Than One Principle of Equality

**Vocabulary**

addition principle • parentheses • simplify • multiplication principle
division principle

1.  The second step in the procedure of solving an equation in the form $ax + b =$ some number is to apply the multiplication or _____ to get the variable $x$ alone on one side of the equation.

2.  The second step in the procedure of solving equations is to _____ each side of the equation, which involves combining like terms and simplifying numerical work.

3.  The first step in the procedure of solving an equation in the form $ax + b =$ some number is use the _____ to get the variable term $ax$ alone on one side of the equation.

4.  The first step in the procedure to solve equations is removing any _____.

| Example | Student Practice |
|---|---|
| **1.** Solve $-6x - 4 = 74$. | **2.** Solve $-5a - 20 = 120$. |

**Example**

**1.** Solve $-6x - 4 = 74$.

First, we must get the variable term, $-6x$, alone by using the addition principle. Add 4 to both sides of the equation.

$$-6x - 4 = 74$$
$$\underline{+\qquad 4\ +4}$$
$$-6x \qquad = 78$$

Then apply the division principle to get the variable, $x$, alone. Divide by $-6$ on both sides of the equation.

$$-6x = 78$$
$$\frac{-6x}{-6} = \frac{78}{-6}$$
$$x = -13$$

Vocabulary Answers: 1. division principle  2. simplify  3. addition principle  4. parentheses

| Example | Student Practice |
|---|---|
| **3.** Solve $-3 = 6 - 4y$ and check your solution. | **4.** Solve $-4 = 6 - 7x$ and check your solution. |

First, use the addition principle to get the variable term, $-4y$, alone. Add $-6$ to both sides of the equation.

$$-3 = -6 - 4y$$
$$\underline{+\ -6 = -6\phantom{-4y}}$$
$$-9 = \phantom{-}-4y$$

Now use the division principle to get $y$ alone on the right side of the equation. Divide by $-4$ on both sides of the equation.

$$\frac{-9}{-4} = \frac{-4y}{-4}$$

$$\frac{9}{4} = y$$

Check the solution by replacing $y$ with $\frac{9}{4}$ and verifying that a true statement results.

$$-3 = 6 - 4y$$
$$-3 \overset{?}{=} 6 - 4\left(\frac{9}{4}\right)$$
$$-3 \overset{?}{=} 6 - 9$$
$$-3 = -3$$

Since $-3 = -3$ is true, the solution is $y = -\dfrac{9}{4}$.

| Example | Student Practice |
|---|---|
| **5.** Solve $5x - 3 = 6x + 2$ and check your solution. | **6.** Solve $7x - 3 = 8x + 5$ and check your solution. |

First, add $-6x$ to both sides of the equation so that all variable terms are on one side of the equation.

$$
\begin{array}{rl}
5x - 3 = & 6x + 2 \\
+ \ -6x \quad = & -6x \\
\hline
-x - 3 = & 2
\end{array}
$$

Then, solve the equation. Add 3 to both sides of the equation.

$$
\begin{array}{rl}
-x - 3 = & 2 \\
+ \quad +3 & +3 \\
\hline
-x \quad = & 5
\end{array}
$$

Now write $-x$ as $-1x$, and divide by $-1$ on both sides.

$$\frac{-1x}{-1} = \frac{5}{-1}$$
$$x = -5$$

Another way to solve $-1x = 5$ is to multiply both sides of the equation by $-1$.

$$-1x = 5$$
$$-1(-1x) = -1(5)$$
$$x = 5$$

Check by replacing $x$ with $-5$ in the original equation and verify that a true statement results.

**Extra Practice**

**1.** Solve and check your solution.

$$8x + 1 - 3x - 7 = 7$$

**2.** Solve and check your solution.

$$-7 + 12x - 12 = 7x - 3$$

**3.** Solve and check your solution.

$$-3x + 5 + 4x - 3 = 15 - 27$$

**4.** Solve and check your solution.

$$-9x + 5 + 7x = -3x + 10$$

**Concept Check**

Explain the steps to solve the equation $8x + 3 - 4x = 15$.

Name: _____     Date: _____

Instructor: _____     Section: _____

**Chapter 7 Solving Equations**
**7.3 Solving Equations with Parentheses**

**Vocabulary**
parentheses   •   simplify   •   isolate the $ax$ term   •   isolate the $x$   •   check
sign   •   opposite

1. When simplifying and solving equations, you have to pay special attention to the
   _____ of each term.

2. The second step in the procedure to solve equations is to _____ each side of the
   equation, which involves combining like terms and simplifying numerical work.

3. When a negative sign is in front of parentheses, we find the _____ of the
   expression by changing the sign of each term inside the parentheses.

4. The first step in the procedure to solve equations is to remove any _____.

| Example | Student Practice |
|---|---|
| **1.** Solve $-3(2x+1)+4x = 27$ and check your solution. | **2.** Solve $-5(4x+2)+6x = 74$ and check your solution. |

Use the distributive property to remove parentheses and then simplify and solve.

$$-3(2x+1)+4x = 27$$
$$-6x-1+4x = 27$$
$$-2x-3 = 27$$
$$\underline{+\quad +3\ +3}$$
$$-2x\quad = 30$$
$$\frac{-2x}{-2} = \frac{30}{-2}$$
$$x = -15$$

Check by replacing $x$ with $-15$ in the original equation and verifying that a true statement results.

Vocabulary Answers: 1. sign  2. simplify  3. opposite  4. parentheses

| Example | Student Practice |
|---|---|

**3.** Solve $5(y+8) = -6(y-2) + 94$ and check your solution.

First remove parentheses and simplify each side of the equation separately.

$$5(y+8) = -6(y-2) + 94$$
$$5 \cdot y + 5 \cdot 8 = -6 \cdot y - 6(-2) + 94$$
$$5y + 40 = -6y + 12 + 94$$
$$5y + 40 = -6y + 106$$

Next, add $6y$ to both sides so that all $y$ terms are on one side of the equation.

$$
\begin{array}{rcrcr}
5y & + & 40 & = & -6y & + & 106 \\
+\ 6y & & & & +6y \\
\hline
11y & + & 40 & = & & & 106
\end{array}
$$

Continue to use the principles of equality to solve for $y$.

$$
\begin{array}{rcrcr}
11y & + & 40 & = & 106 \\
+ & & -40 & & -40 \\
\hline
11y & & & = & 66
\end{array}
$$
$$\frac{11y}{11} = \frac{66}{11}$$
$$y = 6$$

Check by replacing 6 for $y$ in the original equation and verifying that a true statement results.

$$5(6+8) \overset{?}{=} -6(6-2) + 94$$
$$5(14) \overset{?}{=} -6(4) + 94$$
$$70 \overset{?}{=} -24 + 94$$
$$70 = 70$$

**4.** Solve $7(x+6) = -4(x-2) - 43$ and check your solution.

| Example | Student Practice |
|---|---|
| **5.** Solve $(2x^2 + 3x + 1) - (2x^2 + 6) = 4x + 1$. | **6.** Solve $(4x^2 + 5x + 3) - (4x^2 + 8) = 6x + 3$. |

**Example:**

Since there is a minus sign in front of the parentheses, $-(2x^2 + 6)$, we change the sign of each term inside the parentheses and remove the parentheses. Then we simplify the left side by combining like terms.

$$(2x^2 + 3x + 1) - (2x^2 + 6) = 4x + 1$$
$$2x^2 + 3x + 1 - 2x^2 - 6 = 4x + 1$$
$$3x - 5 = 4x + 1$$

Finally, solve for $x$.

$$
\begin{array}{rcl}
3x - 5 &=& 4x + 1 \\
+\ -3x & & -3x \\
\hline
-5 &=& x + 1 \\
+\ -1 & & -1 \\
\hline
-6 &=& x
\end{array}
$$

Check by replacing $x$ with $-6$ in the original equation and verify that the true statement results.

**Extra Practice**

**1.** Solve and check your solution.

$$-5(2x+1)+4x=7$$

**2.** Solve and check your solution.

$$(2y^2-4y+1)-(2y^2+4)=29$$

**3.** Solve and check your solution.

$$(6x^2-3x+1)-(6x^2-6)=2x+12$$

**4.** Solve and check your solution.

$$(y^2+3y-4)-(y^2+4y-6)=6$$

**Concept Check**

**(a)** The first step we should perform to solve $2(y-1)+2=-3(y-2)$ is to simplify the equation using the _____ property.

**(b)** Complete this simplification, then explain the rest of the steps needed to solve the equation so that the variable is on the left side of the equation: $y=\text{some number}$.

**Chapter 7 Solving Equations**
**7.4 Solving Equations with Fractions**

**Vocabulary**
clearing the fractions • least common denominator • every term on both sides • check

1. We multiply all terms on both sides of the equation by the _____ of all the fractions contained in the equation.

2. It is important to multiply _____ of the equation by the least common denominator.

3. The last step in the procedure to solve an equation containing fractions is to _____ your solution.

4. The process of transforming the given equation containing fractions to an equivalent equation that does not contain fractions to avoid unnecessary work is called _____.

| **Example** | **Student Practice** |
|---|---|
| **1.** Solve $\dfrac{x}{3}+\dfrac{x}{2}=5$ and check your solution. | **2.** Solve $\dfrac{x}{4}+\dfrac{x}{5}=9$ and check your solution. |

First, clear the fractions from the equation by multiplying each term by the $LCD = 6$ ; then solve the equation.

$$6\left(\frac{x}{3}\right)+6\left(\frac{x}{2}\right)=6(5)$$
$$2x+3x=30$$
$$5x=30$$
$$\frac{5x}{5}=\frac{30}{5}$$
$$x=6$$

Check by replacing $x$ with 6 in the original equation.

Vocabulary Answers: 1. least common denominator 2. every term on both sides 3. check
4. clearing the fractions

| Example | Student Practice |
|---|---|

**3.** Solve $-4x + \dfrac{3}{2} = \dfrac{2}{5}$.

**4.** Solve $-7x + \dfrac{4}{5} = \dfrac{9}{4}$.

The LCD is 10. We multiply each term by 10.

$$-4x + \frac{3}{2} = \frac{2}{5}$$

$$10(-4x) + 10\left(\frac{3}{2}\right) = 10\left(\frac{2}{5}\right)$$

$$-40x + 15 = 4$$

Now solve for $x$.

$$
\begin{array}{rrrr}
-40x & + & 15 & = & 4 \\
+ & & -15 & & -15 \\
\hline
-40x & & & = & -11
\end{array}
$$

$$\frac{-40x}{-40} = \frac{-11}{-40}$$

$$x = \frac{11}{40}$$

Check the solution by replacing $x$ with $\dfrac{11}{40}$ in the original equation and verify that a true statement results.

$$-4\left(\frac{11}{40}\right) + \frac{3}{2} \overset{?}{=} \frac{2}{5}$$

$$-\frac{11}{10} + \frac{3}{2} \overset{?}{=} \frac{2}{5}$$

$$10\left(-\frac{11}{1}\right) + 10\left(\frac{3}{2}\right) \overset{?}{=} 10\left(\frac{2}{5}\right)$$

$$-11 + 15 \overset{?}{=} 4$$

$$4 = 4$$

| Example | Student Practice |
|---|---|
| **5.** Solve $\dfrac{x}{7} + x = 8$. | **6.** Solve $\dfrac{x}{5} + x = 6$. |

There is only one denominator; thus the LCD is 7.

Multiply each term by the LCD, 7.

$$\frac{x}{7} + x = 8$$

$$7\left(\frac{x}{7}\right) + 7(x) = 7(8)$$

$$1x + 7x = 56$$

Now combine like terms and solve.

$$1x + 7x = 56$$
$$8x = 56$$
$$\frac{8x}{8} = \frac{56}{8}$$
$$x = 7$$

Check the solution by replacing x with 7 in the original equation and verifying that a true statement results.

$$\frac{x}{7} + x \stackrel{?}{=} 8$$

$$\frac{7}{7} + 7 \stackrel{?}{=} 8$$

$$1 + 7 \stackrel{?}{=} 8$$

$$8 = 8$$

Since a true statement results, the solution to the equation is $x = 7$.

**Extra Practice**

1. Solve $6x + \dfrac{1}{2} = \dfrac{2}{3}$. Check your solution.

2. Solve $-4x - \dfrac{2}{3} = \dfrac{5}{6}$. Check your solution.

3. Solve $\dfrac{x}{4} - 5x = 19$. Check your solution.

4. Solve $\dfrac{x}{10} + x = 7$. Check your solution.

**Concept Check**

Explain the steps to solve the equation $-2x + \dfrac{3}{4} = \dfrac{1}{2}$.

**Chapter 7 Solving Equations**
**7.5 Using Equations to Solve Applied Problems**

**Vocabulary**

perimeter      •      area      •      check      •      variable expression

1.  For a rectangle, we can use the formula $A = LW$ to find the _____.

2.  When an applied problem involves comparing two or more quantities we must write a _____ that describes one quantity in terms of another.

3.  The last step in the procedure to solve applied problems is to _____ the solution.

4.  To find the _____ of a shape, we find the sum of all the sides.

| **Example** | **Student Practice** |
|---|---|
| **1.** Art has a rectangular planter box in his front yard that has width $= 2$ ft and length $= 6$ ft . Art plans to increase the length by $x$ ft so that the new perimeter is 32 ft. How much should the length be enlarged? | **2.** Refer to example **1** to answer the following. Art increases the length by $x$ ft so that the new perimeter is 48 ft. |
| | **(a)** How much should the length be enlarged? |

$L = 6 \text{ ft} + x$

$W = 2 \text{ ft}$ | $P = 32 \text{ ft}$ |

Read the problem carefully and create a Mathematics Blueprint.

Write the formula for the perimeter and replace $P$, $L$, and $W$ with the values given. Then solve for $x$.

**(b)** What will the length of the enlarged planter box be?

$$P = 2L + 2W$$
$$32 = 2(x+6) + 2(2)$$
$$32 = 2x + 12 + 4$$
$$16 = 2x$$
$$8 = x$$

The length should be enlarged 8 ft.

Vocabulary Answers: 1. area  2. variable expression  3. check  4. perimeter

| Example | Student Practice |
|---|---|
| **3.** Linda is a store manager. The assistant manager, Erin, earns $8500 less annually than Linda does. The sum of Linda's annual salary and Erin's annual salary is $72,000. How much does each earn annually? | **4.** Jack is a foreman for a construction company. The apprentice for the company earns $6500 less annually than Jack does. The sum of Jack's annual salary and the apprentice's annual salary is $75,000. How much does each earn annually? |

**3.** (continued)

**(a)** Define the variable expressions.

Read the problem carefully and create a Mathematics Blueprint.

Since we are comparing Erin's salary to Linda's, we let the variable $L$ represent Linda's salary.

Linda's salary $= L$
Erin's salary $= (L - 8500)$

**(a)** Define the variable expressions.

**(b)** Write an equation.

Linda's salary $+$ Erin's salary $=$ total annual salary for both people
$L + (L - 8500) = 72,000$

**(b)** Write an equation.

**(c)** Solve the equation and determine the values asked for.

$$L + (L - 8500) = 72,000$$
$$2L - 8500 = 72,000$$
$$\underline{+\quad 8500 \qquad 8,500}$$
$$2L = 80,500$$
$$L = 40,250$$

Linda earns $40,250 annually.

$$L - \$8500 = \text{Erin's salary}$$
$$\$40,250 - \$8500 = \$31,750$$

Erin earns $31,750 annually.

**(c)** Solve the equation and determine the values asked for.

## Extra Practice

**1.** Find $x$ if the perimeter of the rectangle is 56 meters.

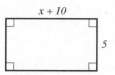

**2.** Find $x$ if the perimeter of the triangle is 22 inches.

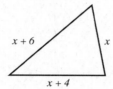

**3.** Bob, an experienced assistant store manager, earns $2500 more annually than Carl, a new assistant store manager. The sum of Bob's annual salary and Carl's annual salary is $41,500. How much does each earn annually?

**4.** Sue, an executive, earns twice as much as her secretary Sam annually. The sum of Sue's annual salary and Sam's annual salary is $93,000. How much does each earn annually?

## Concept Check

The high school marching band fundraiser for uniforms requires that students find donors to pledge money based on the number of laps they run on the school track field. This year Miguel ran 2 more laps than Sam, and Alicia ran 4 fewer laps than Sam. The total laps completed by all three was 29.

Eduardo wanted to calculate the number of laps Alicia completed. To find this information, he let $S$ represent the number of laps that Sam ran, and then solved the equation $(S+2)+(4-S)=29$ to find the number of laps completed by Alicia.

(a) Did Eduardo use the correct equation? Why or why not?

(b) How many laps did Miguel complete for the fundraiser?

# MATH COACH

*Mastering the skills you need to do well on the test.*

Watch the **MATH COACH** videos in MyMathLab® or on YouTube™ while you work the problems below. These helpful hints will help you avoid making common errors on test problems.

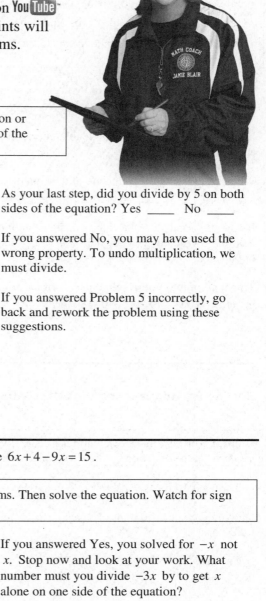

### Solving Equations Using More that One Principle of Equality—Problem 5  Solve $5x - 11 = -1$.

> **Helpful Hint:** Use the addition principle *before* the multiplication or division principle. Remember that whatever you do to one side of the equation, you must do to the other side of the equation.

Examine your work for Problem 5.

As your first step, did you use the division principle and divide by 5? Yes _____ No _____

If you answered Yes, your resulting equation contains fractions. To simplify calculations, use the addition principle first. Please stop now and complete this step.

Did you add 11 to *both sides* of the equation?
Yes _____ No _____

If you answered No, reread the Helpful Hint box and complete this step again.

As your last step, did you divide by 5 on both sides of the equation? Yes _____ No _____

If you answered No, you may have used the wrong property. To undo multiplication, we must divide.

If you answered Problem 5 incorrectly, go back and rework the problem using these suggestions.

---

### Simplifying and Solving Equations—Problem 10  Solve $6x + 4 - 9x = 15$.

> **Helpful Hint:** First simplify the equation by combining like terms. Then solve the equation. Watch for sign errors.

Did you get $-3x + 4 = 15$ when you combined like terms?
Yes _____ No _____

Next did you subtract 4 from both sides of the equation?
Yes _____ No _____

If you answered No to either question, go back and complete these steps again.

As your last step, did you divide both sides by 3?
Yes _____ No _____

If you answered Yes, you solved for $-x$ not $x$. Stop now and look at your work. What number must you divide $-3x$ by to get $x$ alone on one side of the equation?

Now go back and rework the problem using these suggestions.

**Solving Equations with Parentheses—Problem 14**    Solve $4(x-1) = -6(x+2)+48$ .

---

**Helpful Hint:**

- Did you remove parentheses?
- Did you simplify each side of the equation separately?
- Did you isolate $x$ terms on one side of the equation and then use the proper principle of equality to solve for $x$ ?

---

After removing parentheses, did you get $4x - 4 = -6 - 12 + 48$ ? Yes _____    No _____

If you answered No, stop and double-check all your calculations.

When simplifying the right side of the equation, did you get $-6x + 36$ ? Yes _____    No _____

If you answered No, take care with signs. You may find it helpful to write out all your steps and keep your work organized.

To get all of the $x$ terms on one side of the equation, did you get either $10x = 40$, or $-40 = -10x$ ? Yes _____    No _____

If you answered No, examine your work carefully. Try to identify your error before finishing the problem.

If you answered Problem 14 incorrectly, go back and rework the problem using these suggestions.

---

**Solving Equations Using the LCD Method—Problem 16**    Solve $\dfrac{x}{5} + \dfrac{x}{2} = 7$ .

---

**Helpful Hint:** To make calculations easier, always clear the fractions first. Be careful not to forget to multiply *all terms* in the equation by the least common denominator (LCD).

---

Did you get 10 for the LCD? Yes _____    No _____

Did you remember to multiply each term, including 7, by the LCD? Yes _____    No _____

If you answered No to either question, stop and make these corrections.

Did you get $7x = 70$ after clearing fractions and simplifying? Yes _____    No _____

If you answered No, examine your work closely. Check each step of the process for a calculation error.

Now go back and rework the problem using these suggestions.

## Chapter 8 Decimals and Percents
## 8.1 Understanding Decimal Fractions

**Vocabulary**

decimal point     •     place-value chart     •     decimal fraction     •     rounding

1. The "." in the decimal .3 is called a _____.

2. The rule for _____ decimals is similar to the rule for whole numbers.

3. A _____ is helpful in understanding the meaning of decimal fractions as well as how to write decimal fractions in different forms.

4. A _____ is a fraction whose denominator is a power of 10.

| Example | Student Practice |
|---|---|
| **1.** Write a word name for the decimal 0.561.<br><br>The place value of the last digit is the last word in the word name. The last digit is 1 and is in the thousandths place. We do not include 0 as part of the word name.<br><br>thousandths place<br>$\downarrow$<br>0.56$\boxed{1}$<br><br>Five hundred sixty-one thousandths | **2.** Write a word name for each decimal.<br><br>**(a)** 0.734<br><br><br><br>**(b)** 6.23 |
| **3.** Write the word name for a check written to Shandell Strong for $126.87.<br><br>One hundred twenty-six and $\dfrac{87}{100}$ Dollars | **4.** Write the word name for a check written to Spin-to-Win Sprinklers for $343.51.<br><br>_____ Dollars |

Vocabulary Answers: 1. decimal point  2. rounding  3. place-value chart  4. decimal fraction

| Example | Student Practice |
|---|---|
| **5.** Write 0.86132 using fractional notation. Do not simplify. | **6.** Write 1.4928 using fractional notation. Do not simplify. |

**5.** We do not need to write 0 as part of the fraction.

$$0.86132 = \frac{86,132}{100,000}$$

Notice that the decimal has 5 decimal places and the power of ten in the denominator has 5 zeros.

---

**7.** Write $7\frac{56}{1000}$ as a decimal.

Since there are 3 zeros, we move the decimal point 3 places to the left.

$$7\frac{56}{1000} = 7\frac{\overleftarrow{056}}{1000} = 7.056$$

Note that we had to insert a 0 before 56 so we could move the decimal point 3 places to the left.

**8.** Write $1\frac{23}{1000}$ as a decimal.

---

**9.** Replace the ? with < or >. 0.24 ? 0.244

Add a zero to 0.24 so that both decimal parts have the same number of digits.

0.240 ? 0.244

The tenths digits and the hundredths digits are equal. The thousandths digits differ.

Since $0 < 4$, $0.240 < 0.244$.

**10.** Replace the ? with < or >. 0.236 ? 0.23

| Example | Student Practice |
|---|---|
| **11.** Round 237.8435 to the nearest hundredth.<br><br>The round-off place digit is in the hundredths place.<br><br>204.8⬚4⬚35<br><br>The digit to the right of the round-off place digit is less than 5, so we do not change the round-off place digit. Drop all digits to the right of the round-off place digit.<br><br>237.84<br><br>237.8435 rounded to the nearest hundredth is 237.84. | **12.** Round 34.05482 to the nearest thousandth. |
| **13.** Round to the nearest hundredth. Alex and Lisa used 204.9954 kilowatt-hours of electricity in their house in June.<br><br>We locate the digit in the hundredths place.<br><br>204.9⬚9⬚54<br><br>Since the digit to the right of 9 is 5, we increase 9 to 10 by changing 9 to 0. Then we must increase the 9 in the tenths place by 1, followed by increasing 4 to 5.<br><br>205.00<br><br>Note that we must include the two zeros after the decimal point because we were asked to round to the nearest hundredth.<br><br>Thus, 204.9954 rounds to 205.00 kilowatt-hours. | **14.** Round to the nearest hundredth. In November, Cassie and Roger used 150.9971 kilowatt-hours of electricity in their house. |

181

**Extra Practice**

**1.** Write a word name for the decimal 3.48.

**2.** Write the fraction $19\frac{243}{1000}$ as a decimal.

**3.** Replace the ? with $<$ or $>$. 0.35 ? 0.355

**4.** Round 3.1415 to the nearest thousandth.

**Concept Check**

Explain how you know how many zeros to put in the denominator of your answer when you write 8.6711 as a fraction.

Name: _____     Date: _____

Instructor: _____     Section: _____

## Chapter 8 Decimals and Percents
## 8.2 Adding and Subtracting Decimal Expressions

**Vocabulary**

line up    •    evaluate    •    combine    •    estimate

1. To _____ an expression, replace the variable with the given number and simplify.

2. The first step to add or subtract decimals is to write the numbers vertically and _____ the decimal points.

3. We can _____ a sum or difference of decimals by rounding each decimal to the nearest whole number.

4. To _____ like terms we add coefficients of the like terms and the variable part stays the same.

| Example | Student Practice |
|---|---|
| **1.** Add $40 + 8.77 + 0.9$. | **2.** Add $28 + 3.87 + 0.2$. |
| We can write any whole number as a decimal by placing a decimal point at the end of the number: $40 = 40$. Line up the decimal points and add zeros so that each number has the same number of decimal places. | |

$$\begin{array}{r} \overset{1}{4}0.00 \\ 8.77 \\ + 0.90 \\ \hline 49.67 \end{array}$$

| **3.** Subtract $19.02 - 8.6$. | **4.** Subtract $32.06 - 4.2$. |
|---|---|

$$\begin{array}{r} 1\overset{8}{\cancel{9}}.\overset{10}{\cancel{0}}2 \\ -8.6\,0 \\ \hline 10.4\,2 \end{array}$$

Vocabulary Answers: 1. evaluate  2. line up  3. estimate  4. combine

| Example | Student Practice |
|---|---|
| **5.** Perform the operation indicated.<br><br>$-9.79 - (-0.68)$<br><br>To subtract, we add the opposite of the second number.<br><br>$-9.79 - (-0.68)$<br><br>$-9.79 + (0.68)$<br><br>Next, to add numbers with different signs, we keep the sign of the larger absolute value and subtract.<br><br>$\quad -9.79$<br>$\quad \underline{\phantom{-}0.68}$<br>$\quad -9.11$<br><br>The answer is negative since $\lvert -9.79 \rvert$ is larger than $\lvert 0.68 \rvert$. | **6.** Perform the operation indicated.<br><br>$-2.64 - (-10.23)$ |
| **7.** Combine like terms. $11.2x + 3.6x - 7.1y$<br><br>$11.2x$ and $3.6x$ are like terms, so we line up the decimal points and add them.<br><br>$\quad 11.2x$<br>$\underline{+\ 3.6x}$<br>$\quad 14.8x$<br><br>We have $11.2x + 3.6x - 7.1y$<br>$\qquad = 14.8x - 7.1y.$<br><br>We cannot combine $14.8x$ and $7.1y$ since they are not like terms. | **8.** Combine like terms. $2.3b + 7.4b - 8.2a$ |

| Example | Student Practice |
|---|---|
| **9.** Evaluate $x + 3.12$ for $x = 0.11$. | **10.** Evaluate $y + 5.89$ for $y = -7.21$. |

**9.** Evaluate $x + 3.12$ for $x = 0.11$.

We replace the variable with 0.11.

$x + 3.12 = 0.11 + 3.12$

Next we line up decimal points and add.

$$\begin{array}{r} 3.12 \\ + 0.11 \\ \hline 3.23 \end{array}$$

**11.** Julie runs on her treadmill every day, She wants to run approximately 25 miles each week to prepare for a track race. She logged the distance she ran each day this week on the following chart. Estimate the total number of miles Julie ran this week.

| Monday | Tuesday | Wednesday | Thursday |
|---|---|---|---|
| 2.13 mi | 2.79 mi | 2.9 mi | 3.11 mi |

| Friday | Saturday | Sunday | |
|---|---|---|---|
| 3.8 mi | 4.12 mi | 4.9 mi | |

We round each decimal to the nearest whole number and then add.

$2 + 3 + 3 + 3 + 4 + 4 + 5 = 24$ miles

Julie ran approximately 24 miles.

**12.** Antonio has kept track of how much he spends at the Laundromat for his last 5 trips. The total amount of each trip is as follows: $9.75, $11.25, $10.50, $10.75, $8.75. Estimate the total amount he spent at the Laundromat over the past 5 trips.

| | | Example | | | Student Practice |

**Example**

13. The table compares some of the top players' average season statistics.

| MVP | Points | Rebounds |
|---|---|---|
| 1. Allen Iverson | 31.1 | 3.1 |
| 2. Tim Duncan | 25.5 | 12.7 |
| 3. Kobe Bryant | 28.3 | 6.3 |
| 4. LeBron James | 29.7 | 7.3 |

Source: Orange County Register

(a) How many more points did Iverson average than Bryant?

$$
\begin{array}{r}
\overset{10}{\cancel{3}} \overset{2\ \cancel{0}\ 11}{\cancel{1}.\cancel{1}} \\
\text{Allen Iverson's points} \\
\text{Kobe Bryant's points} \quad -28.3 \\
\hline
2.8
\end{array}
$$

(b) Find the total of the average number of rebounds made by Iverson, Duncan, and James.

$$
\begin{array}{r}
\text{Iverson} \quad 3.1 \\
\text{Duncan} \quad 12.7 \\
\text{James} \quad 7.3 \\
\hline
23.1
\end{array}
$$

**Student Practice**

14. Use the table in example 13 to answer the following.
    (a) How many more rebounds did Duncan average than James?

(b) Find the total of the average number of points made by all four players listed on the chart.

**Extra Practice**

1. Add $7.334 + 21.04$.

2. Subtract $-12.6 - (-6.2)$.

3. Combine like terms. $10.02x - 7.39x + 3.4y$

4. Evaluate $x - 0.97$ for $x = 13.23$.

**Concept Check**

Explain how you would evaluate $x - 3.1$ for $x = 0.866$.

**Chapter 8 Decimals and Percents**
**8.3 Multiplying and Dividing Decimal Expressions**

**Vocabulary**
repeating decimals   •   positive   •   negative   •   dividend

1. To divide a decimal by a whole number, we place the decimal point in the quotient directly above the decimal point in the _____.

2. Decimals that have a digit, or group of digits, that repeats are called _____.

3. To multiply or divide positive and negative decimals, the sign of the answer will be _____ if the problem has an odd number of negative signs.

4. To multiply or divide positive and negative decimals, the sign of the answer will be _____ if the problem has an even number of negative signs.

| Example | Student Practice |
|---|---|
| **1.** Multiply $5.33 \times 7.2$.<br><br>We write the multiplication just as we would if there were no decimal points. We need 3 decimal places since there are 2 in the first factor and 1 in the second.<br><br>$\quad\ 5.33$<br>$\underline{\times\, 7.2}$<br>$\quad 1066$<br>$\underline{3731}$<br>$\ 38.376$ | **2.** Multiply $34.1 \times 2.78$. |
| **3.** Multiply $(-2)(4.51)$.<br><br>The number of negative signs, 1, is odd so the product is negative.<br><br>$\quad\ 4.51$<br>$\underline{\times (-2)}$<br>$\ -9.02$ | **4.** Multiply $(-5)(7.93)$. |

Vocabulary Answers: 1. dividend  2. repeating decimals  3. negative  4. positive

| Example | Student Practice |
|---|---|
| **5.** Multiply $0.2345 \times 1000$.<br><br>Since 1000 has three zeros, we move the decimal point to the right three places.<br><br>$0.2345 \times 1000 = 234.5$ | **6.** Multiply $(0.4256)(10^6)$. |
| **7.** Divide $2.3 \div 5$.<br><br>Place the decimal point directly above the decimal point in the dividend and divide as if there were no decimal point. Add zeros as needed.<br><br>$$2.3 \div 5 \rightarrow 5\overline{)\begin{array}{l}0.46 \\ 2.30\end{array}}$$<br>$$\begin{array}{r} 2\,0 \\ \hline 30 \\ 30 \\ \hline 0 \end{array}$$ | **8.** Divide $3.6 \div 8$. |
| **9.** Divide $-0.185 \div 13$. Round your answer to the nearest thousandth.<br><br>We must divide one place beyond the thousandths place, to the ten thousandths place, so we can round to the nearest thousandth.<br><br>$$13\overline{)\begin{array}{l}-.0142 \\ -0.1850\end{array}}$$<br>$$\begin{array}{r} 13 \\ \hline 55 \\ 52 \\ \hline 30 \\ 26 \\ \hline 4 \end{array}$$<br>$-0.185 \div 13 \approx -0.014$ | **10.** Divide $-0.8469 \div 17$. Round your answer to the nearest thousandth. |

| Example | Student Practice |
|---|---|
| **11.** Divide $6.93 \div 2.2$. | **12.** Divide $10.92 \div 2.4$. |

**11.** Divide $6.93 \div 2.2$.

Since the divisor, 2.2, is not a whole number, multiply the dividend and quotient by 10 and the divisor becomes a whole number.

$$\frac{(6.93)(10)}{(2.2)(10)} = \frac{69.3}{22} = 69.3 \div 22$$

$$\begin{array}{r} 3.15 \\ 22{\overline{\smash{\big)}\,69.30}} \\ \underline{66\phantom{.30}} \\ 33\phantom{.0} \\ \underline{22\phantom{.0}} \\ 110 \\ \underline{110} \\ 0 \end{array}$$

$6.93 \div 2.2 = 3.15$

**13.** Divide $0.7 \div 1.5$.

The divisor is not a whole number, so move the decimal point one place to the right.

$$\begin{array}{r} 0.466 \\ 15{\overline{\smash{\big)}\,7.000}} \\ \underline{6\,0\phantom{00}} \\ 1\,00\phantom{0} \\ \underline{90\phantom{0}} \\ 100 \\ \underline{90} \\ 10 \end{array}$$

Thus, $0.7 \div 1.5 = 0.4\overline{6}$ because if we continued dividing, the 6 would repeat.

**14.** Divide $1.3 \div 1.6$.

| Example | Student Practice |
|---|---|
| **15.** Write as a decimal. $5\frac{7}{11}$ | **16.** Write as a decimal. $3\frac{7}{12}$ |

$5\frac{7}{11}$ means $5+\frac{7}{11}$, so divide 7 by 11.

$$
\begin{array}{r}
0.6363 \\
11\overline{)7.0000} \\
\underline{6\,6}\phantom{000} \\
40\phantom{00} \\
\underline{33}\phantom{00} \\
70\phantom{0} \\
\underline{66}\phantom{0} \\
40 \\
\underline{33}
\end{array}
$$

Since the pattern of 40 minus 33 repeats, $\frac{7}{11}=0.\overline{63}$ and $5\frac{7}{11}=5.\overline{63}$.

## Extra Practice

**1.** Multiply $(15.23)(-3)$.

**2.** Multiply $2.3681\times1000$.

**3.** Divide $13.5525\div4.17$.

**4.** Write $11\frac{4}{15}$ as a decimal. Round to the nearest hundredth when necessary. If a repeating decimal is obtained, use proper notation such as $0.\overline{3}$.

## Concept Check

Marc multiplied $0.097\times0.5$ and obtained the answer 0.485. Is Marc's answer correct? Why or why not?

**Chapter 8 Decimals and Percents**
**8.4 Solving Equations and Applied Problems Involving Decimals**

**Vocabulary**
power of 10     •     decimal

1. We often deal with _____ equations when we solve problems involving money.

2. We can eliminate decimal terms in an equation by multiplying both sides of the equation by the _____ necessary to eliminate the decimal from every term.

| Example | Student Practice |
|---|---|
| **1.** Solve $x - 4.51 = 6.74$. <br><br> $x - 4.51 = 6.74$ <br> $+ \quad 4.51 \quad 4.51$ <br> $\quad\quad x = 11.25$ | **2.** Solve $b - 5.63 = 7.12$ and check your solution. |
| **3.** Solve $2(x + 1.5) = x - 4.62$. <br><br> $2(x) + 2(1.5) = x - 4.62$ <br> $2x + 3 = x - 4.62$ <br> $x + 3 = -4.62$ <br> $x = -7.62$ | **4.** Solve $5(x + 1.7) = 4x + 3.2$. |
| **5.** Solve $0.35x + 0.3 = 1.7$. <br><br> Multiply both sides of the equation by 100. <br> $0.35x + 0.3 = 1.7$ <br> $100(0.35x + 0.3) = 100(1.7)$ <br> $100(0.35x) + 100(0.3) = 100(1.7)$ <br> $35x + 30x = 170$ <br> $35x = 140$ <br> $\dfrac{35x}{35} = \dfrac{140}{35}$ <br> $x = 4$ | **6.** Solve $1.1x + 0.32 = 10$. |

Vocabulary Answers: 1. decimal  2. power of 10

| Example | Student Practice |
|---|---|

**7.** A long-distance phone carrier in New Jersey charges a base fee of $4.95 per month, plus 10 cents per minute for calls outside New Jersey and 5 cents per minute for long-distance calls within the state of New Jersey. On average, Natasha's monthly long-distance calls total 45 minutes outside the state and 220 minutes within the state. How much more or less will Natasha's average long-distance bill be than her budget of $25 per month?

**8.** Refer to the information in example **7** to answer the following. Natasha averages a total of 65 minutes of long-distance calls outside the state and a total of 250 minutes within the state. How much more or less will her average monthly bill be than her budget of $25?

Read the problem carefully and create a Mathematics Blueprint.

Average long-distance bill = base fee + charge for calls outside state + charge for calls inside state

$$x = 4.95 + (45 \text{ min} \times 10\text{¢ per min})$$
$$+ (220 \text{ min} \times 5\text{¢ per min})$$
$$= 4.95 + 45 \times 0.10 + 220 \times 0.05$$
$$= 4.95 + 4.50 + 11$$
$$x = 20.45$$

$20.45 is the average cost of Natasha's long-distance calls, and this amount is less than her monthly budget of $25. We subtract: $25 − $20.45 = $4.55. Her average monthly bill will be $4.55 less than her budget of $25.

Check by using your calculator to verify your results.

| Example | Student Practice |
|---|---|

**9.** Andy must fill his vending machine with change. From past experience he knows that the number of dimes he needs to place in the machine for change is twice the number of quarters. The total value of dimes and quarters needed for change is $4.95. How many of each coin must Andy put in the vending machine?

**10.** Maria must fill her vending machine with change. From past experience she knows that the number of dimes she needs to place in the machine for change is three times the number of quarters. The total value of dimes and quarters needed for change is $3.85. How many of each coin must Maria put in the vending machine?

Read the problem carefully and create a Mathematics Blueprint.

We define our variables to find the value of each coin. Since we are comparing dimes to quarters, we let our variable represent the number of quarters.

$Q$ = number of quarters
$2Q$ = number of dimes

We write an equation.

The value of quarters + the value of dimes = the total value of coins

$$(0.25)Q + (0.10)(2Q) = \$4.95$$

We solve the equation and determine the number of dimes and quarters.

$$0.25Q + 0.20Q = 4.95$$
$$25Q + 20Q = 495$$
$$45Q = 495$$
$$\frac{45Q}{45} = \frac{495}{45}$$
$$Q = 11$$
$$2Q = 2(11) = 22$$

Andy must put 11 quarters and 22 dimes in the vending machine for change.

## Extra Practice

**1.** Solve $-2.2x = -13.2$.

**2.** Solve $5(3x - 0.4) = 4x - 2$.

**3.** Solve $5x + 11.5 = x - 21$.

**4.** Joseph went shopping at Sami's CD/DVD Store. He purchased 2 CDs at $13.99 each, 1 CD set for $18.99, and 2 DVDs for $19.99 each (sales tax was included). When reaching the cashier, Joseph had only $100 in his wallet. Did he have enough money?

## Concept Check

Fill in the table below to complete the steps necessary to solve the equation $2.2 + 2.4 = 5(x - 1) + 3x$. In the left column explain how to proceed and in the right column complete the step.

| 1. Add 2.2 + 2.4 | $4.6 = 5(x - 1) + 3x$ |
|---|---|
| 2. | |
| 3. | |
| 4. | |
| 5. | |

**Chapter 8 Decimals and Percents**
**8.5 Estimating with Percents**

**Vocabulary**

percents        •        last digit        •        15%        •        20%

1. To estimate 10% of a whole number, delete the _____.

2. _____ can be described as ratios whose denominators are 100.

3. To estimate _____, double 10%.

4. To estimate _____, add 10% and 5%

---

| **Example** | **Student Practice** |
|---|---|
| **1.** For the number 59,040, estimate the following. | **2.** For the number 74,605, estimate the following. |
| **(a)** 10% | **(a)** 10% |
| To estimate 10% of 59,040, first we round to the nearest thousand, then we delete the last digit. | |
| $59{,}040 \rightarrow 59{,}000$ | |
| $59{,}00\cancel{0}$ | |
| 10% of $59{,}040 \approx 5900$ | |
| **(b)** 1% | **(b)** 1% |
| To estimate 1% of 59,040, first we round to the nearest thousand, then we delete the last two digits. | |
| $59{,}040 \rightarrow 59{,}000$ | |
| $59{,}0\cancel{0}\cancel{0}$ | |
| 1% of $59{,}040 \approx 590$ | |

Vocabulary Answers: 1. last digit  2. percents  3. 20%  4. 15%

| Example | Student Practice |
|---|---|
| **3.** For the number 1205, estimate the following. | **4.** For the number 3206, estimate the following. |

**3.** For the number 1205, estimate the following.

(a) 5% of 1205

We round 1205 to the nearest hundred: 1200.

To find 5%, we find $\frac{1}{2} \times 10\%$ of the number.

$$\frac{1}{2} \times (10\% \text{ of } 1200) = \frac{1}{2} \times 120 = 60$$

(b) 15% of 1205

To find 15%, we add $(10\% \text{ of } 1200) + (5\% \text{ of } 1200)$.

$$15\% \text{ of } 1200 = 120 + 60 = 180$$

(c) 6% of 1205

To find 6% we add: $(5\% \text{ of } 1200) + (1\% \text{ of } 1200)$.

$$6\% \text{ of } 1200 = 60 + 12 = 72$$

**4.** For the number 3206, estimate the following.

(a) 10% of 3206

(b) 20% of 3206

(c) 11% of 3206

---

**5.** Loren would like to leave a 15% tip for her dinner, If the total bill at a restaurant is $20.76, estimate the tip Loren should leave.

We round $20.76 to the nearest ten: $20.

$$15\% \text{ of } 20 = (10\% \text{ of } 20) + (5\% \text{ of } 20)$$
$$= 2 + 1$$
$$= \$3$$

**6.** Doug would like to leave a 20% tip for his breakfast. If the total bill at a restaurant is $11.04, estimate the tip that Doug should leave.

| Example | Student Practice |
|---|---|
| **7.** Everything in a store is on sale for 30% off the original price. The discount is calculated at the cash register at the time of purchase. Josh buys 2 shirts priced at $19.95 each, one pair of pants priced at $28.00, and a pair of shoes priced at $39.99 | **8.** Josh bought 3 shirts, 3 pairs of pants, and 2 pairs of shoes from the same store in example **7.** |

**(a)** Round the original price of each item to the nearest ten and then estimate the total cost of the items before the discount.

Shirt: $19.95 → $20
Pants: $28 → $30
Shoes: $39.99 → $40

We add these amounts:
$20 + $20 + $30 + $40 = $110$.

**(a)** Round the original price of each item to the nearest ten and then estimate the total cost of the items before the discount.

**(b)** Estimate the amount of the 30% discount. Round the discount to the nearest ten.

10% of $110 = $11$
30% of $110 = $11 + $11 + $11$
$= $33$

The estimated discount rounded to the nearest ten is $30.

**(b)** Estimate the amount of the 30% discount. Round the discount to the nearest ten.

**(c)** Estimate the total cost of the items after the discount is taken.

We subtract the total estimated cost minus the estimated discount.

$110 − $30 = $80$

The estimated cost of the items after the discount is $80.

**(c)** Estimate the total cost of the items after the discount is taken.

**Extra Practice**

**1.** For the number 407, estimate 30%.

**2.** For the number 3006, estimate 8%.

**3.** For the number 420,070, estimate 4%.

**4.** Matt and Andrea paid the realtor 7% commission on the sale price of their home. If they sold their home for $205,000, estimate how much commission was paid to the realtor.

**Concept Check**

We can estimate 35% of 200 by finding 3 times 10% of 200, then adding $\frac{1}{2}$ times 10% of 200. Explain two other ways you can estimate 35% of 200.

**Chapter 8 Decimals and Percents**
**8.6 Percents**

**Vocabulary**
percents    •    left    •    right    •    whole

1.  To write a decimal as a percent, move the decimal point two places to the _____.

2.  _____ can be described as ratios whose denominators are 100.

3.  Decimals, fractions, or percents are used to describe parts of a _____.

4.  To write a percent as a decimal, move the decimal point two places to the _____.

| Example | Student Practice |
|---|---|
| **1.** State using percents. 13 out of 100 radios are defective.<br><br>$$\frac{13}{100} = 13\%$$<br><br>13% of the radios are defective. | **2.** State using percents. 3 out of 100 calculators are defective. |
| **3.** Last year's attendance at a school's winter formal was 100 students. This year the attendance was 121. Write this year's attendance as a percent of last year's.<br><br>We must write this year's attendance $(121)$ as a percent of last year's $(100)$.<br><br>This year's attendance $\rightarrow$<br>Last year's attendance $\rightarrow$ $\frac{121}{100} = 121\%$<br><br>This year's attendance at the formal was 121% of last year's.<br><br>Note there is 121 parts out of 100 parts. This means we have more than one whole amount and thus more than 100%. | **4.** Last year, attendance at a town meeting was 100 residents. This year the attendance was 134. Write this year's attendance as a percent of last year's. |

Vocabulary Answers: 1. right  2. percents  3. whole  4. left

| Example | Student Practice |
|---|---|
| **5.** There are 100 milliliters (mL) of a solution in a container, Sara takes 0.3 mL of the solution. What percentage of the solution does Sara take?<br><br>$$\frac{0.3}{100} = 0.3\% \text{ of the solution}$$ | **6.** There are 100 mL of solution in a container. Evan takes 0.5 mL of the solution. What percentage of the solution does Evan take? |

**7.** **(a)** Write 3.8% as a decimal.

We move left on the chart, so the decimal point moves 2 places left.

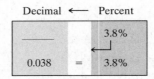

**(b)** Write 0.009 as a percent.

We move right on the chart, so the decimal point moves 2 places right.

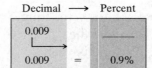

**8.** **(a)** Write 1.5% as a decimal.

**(b)** Write 0.008 as a percent.

**9.** Complete the table of equivalent notations.

| Decimal Form | Percent Form |
|---|---|
| 0.457 | |
| | 58.2% |
| | 0.6% |
| 2.9 | |

| Decimal Form | Percent Form |
|---|---|
| 0.457 | 45.7% |
| 0.582 | 58.2% |
| 0.006 | 0.6% |
| 2.9 | 290% |

**10.** Complete the table of equivalent notations.

| Decimal Form | Percent Form |
|---|---|
| 0.423 | |
| | 64.4% |
| | 0.3% |
| 90.1 | |

| Example | Student Practice |
|---|---|
| **11.** (a) Write $\dfrac{211}{500}$ as a percent. | **12.** (a) Write $\dfrac{67}{125}$ as a percent. |

**11.** (a) Write $\dfrac{211}{500}$ as a percent.

Compute: $211 \div 500 = 0.422$, then move the decimal point 2 places to the right.

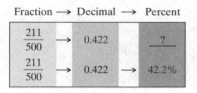

Fraction → Decimal → Percent

**(b)** Write 42.2% as a fraction.

Move the decimal point 2 places to the left. Note that $0.422 = \dfrac{422}{1000} = \dfrac{211}{500}$.

Fraction ← Decimal ← Percent

---

**12.** (a) Write $\dfrac{67}{125}$ as a percent.

**(b)** Write 34.8% as a fraction.

---

**13.** Write $\dfrac{5}{9}$ as a percent. Round to the nearest hundredth of a percent.

First we change $\dfrac{5}{9}$ to a decimal:

$5 \div 9 = 0.55555\ldots$. We must carry out the division at least five places beyond the decimal point so that we can move the decimal point to the right two places, and we then round to the nearest hundredth of a percent.

$0.55555\ldots = 55.555\ldots\% \approx 55.56\%$

Remember that if you are not asked to round, $\dfrac{5}{9} = 55.\overline{5}\%$.

---

**14.** Write $\dfrac{2}{5}\%$ as a fraction.

**Extra Practice**

**1.** Write 245% as a decimal.

**2.** Write 0.07 as a percent.

**3.** Write $5\frac{1}{3}$ as a percent. Round to the nearest hundredth of a percent.

**4.** Write $\frac{1}{8}$% as a fraction.

**Concept Check**

Explain how you would change 0.43% to a decimal, then to a fraction.

**Chapter 8 Decimals and Percents**
**8.7 Solving Percent Problems Using Equations**

**Vocabulary**
base        •        amount        •        percent        •        of

1. We can write the relationship $\dfrac{\text{amount}}{\text{base}} = \text{percent}$ as amount = _____ × base.

2. The part being compared to the base is called the _____.

3. When we translate the statement into symbols, we replace "_____" with ×.

4. The entire quantity is called the _____.

| **Example** | **Student Practice** |
|---|---|
| **1.** Translate into an equation and solve. | **2.** Translate into an equation and solve. |
| **(a)** What is 25% of 40? | **(a)** What is 30% of 30? |
| $n = 25\% \times 40$ | |
| $n = 0.25 \times 40$ | |
| $n = 10$ | |
| **(b)** 10 is 25% of what number? | **(b)** 9 is 30% of what number? |
| $10 = 25\% \times n$ | |
| $10 = 0.25 \times n$ | |
| $\dfrac{10}{0.25} = n$ | |
| $40 = n$ | |
| **(c)** 10 is what percent of 40? | **(c)** 9 is what percent of 30? |
| $10 = n\% \times 40$ | |
| $\dfrac{10}{40} = n\%$ | |
| $0.25 = n\%$ | |
| $25 = n$ | |

Vocabulary Answers: 1. percent  2. amount  3. degrees  4. base

| Example | Student Practice |
|---|---|
| **3.** Translate into an equation and solve. 50 is what percent of 40? | **4.** Translate into an equation and solve. 80 is what percent of 50? |

**Example 3 (continued):**

We should expect to get more than 100% since 50 is more than the base 40.

50 is what percent of 40?

$$50 = n\% \times 40$$

$$\frac{50}{40} = n\%$$

$$1.25 = n\%$$

$$125 = n$$

50 is 125% of 40.

| Example | Student Practice |
|---|---|
| **5.** Find 55% of 36. | **6.** Find 48% of 87. |

**Example 5 (continued):**

$$n = 55\% \times 36$$

$$= 0.55 \times 36$$

$$= 19.8$$

19.8 is 55% of 36.

| Example | Student Practice |
|---|---|
| **7.** Marilyn has 850 out of 1000 points possible in her English class. What percent of the total points does Marilyn have? | **8.** Hank earned 65 out of 80 points on a test. What percentage of total points did he earn? |

**Example 7 (continued):**

We must find the percent, so we write the statement that represents the percent situation.

850 is what percent of 1000?

$$850 = n\% \times 1000$$

$$\frac{850}{1000} = n\%$$

$$0.85 = n\%$$

$$85 = n$$

Marilyn has 85% of the total points.

| Example | Student Practice |
|---|---|
| **9.** Sean's bill for his dinner at the Spaghetti House was $19.75. How much should he leave for a 15% tip? Round this amount to the nearest cent. | **10.** Johanna left a $3 tip for her dinner, which cost $17.89. What percent of the total bill did Johanna leave for a tip? Round your answer to the nearest hundredth of a percent. |

**9.** (continued)

We must find the amount, that is, the part of the base of $19.75.

What is 15% of $19.75?

$n = 15\% \times \$19.75$

$\quad = 0.15 \times \$19.75$

$\quad = \$2.9625$

$\quad \approx \$2.96$

The tip is $2.96

| | |
|---|---|
| **11.** Sergio stayed in a luxury hotel on a Saturday night and paid $230 for that night. If the rate on Saturday night is 15% higher than it is on Sunday night, how much will Sergio pay for the room on Sunday night? | **12.** Refer to example **11** to answer the following. If the rate on Saturday night is 25% higher than it is on Sunday night, how much will Sergio pay to stay Sunday? |

**11.** (continued)

Read the problem carefully and create a Mathematics Blueprint.

Let $x$ = the room rate on Sunday.

The room rate on Sunday plus 15% of the Sunday rate equals the room rate on Saturday.

$100\%x + 15\%x = \$230$

$\qquad 115\%x = 230$

$\qquad 1.15x = 230$

$\qquad x = \dfrac{230}{1.15}$

$\qquad x = 200$

Sergio will pay a rate of $200 to stay Sunday night.

**Extra Practice**

1. Translate into an equation and solve. What is 83% of 155?

2. Translate into an equation and solve. 240 is 80% of what number?

3. Translate into an equation and solve. 280 is what percent of 70?

4. Julie's bill for dinner at the Seafood Hut was $34.75. How much should she leave for a 15% tip? Round the amount to the nearest cent.

**Concept Check**

The owner of M&R Windows determined that 0.8% of the products ordered from the manufacturer are defective. Explain how you would determine how many windows the owner should expect to be defective in a shipment from the manufacturer of 375 windows.

**Chapter 8 Decimals and Percents**
**8.8 Solving Percent Problems Using Proportions**

**Vocabulary**

base • percent • amount • variables

1.  The _____ is the part being compared to the whole.

2.  The letters $a$, $b$, and $p$ to represent amount, base, and percent are called _____.

3.  The _____ is the entire quantity or total involved.

4.  Usually, the easiest part to identify is the _____.

| Example | Student Practice |
|---|---|
| **1.** Identify the percent number $p$. | **2.** Identify the percent number $p$. |
| **(a)** Find 15% of 360. | **(a)** Find 20% of 369. |
| The value of $p$ is 15. | |
| **(b)** 28% of what is 25? | **(b)** 34% of what is 54? |
| The value of $p$ is 28. | |
| **(c)** What percent of 18 is 4.5? | **(c)** What percent of 15 is 3.5? |
| The value of $p$ is unknown. | |
| **3.** Identify the base $b$ and the amount $a$. | **4.** Identify the base $b$ and the amount $a$. |
| **(a)** 25% of 520 is 130. | **(a)** 20% of 42 is 8.4. |
| The base is the entire quantity. $b = 520$. The amount is the part compared to the whole, $a = 130$. | |
| **(b)** 19 is 50% of what? | |
| The amount 19 is the part of the base. The base is unknown. | **(b)** 35 is 40% of what? |

Vocabulary Answers: 1. amount  2. variables  3. base  4. percent

| Example | Student Practice |
|---|---|

**5.** Find the percent $p$, base $b$, and amount $a$.

   **(a)** What is 77% of 210?

   The amount is unknown. The value of $p$ is 77. The base usually follows the word "of." Here, $b = 210$.

   **(b)** What percent of 21 is 17?

   The value of $p$ is not known. The base usually follows the word "of." Here, $b = 21$. The amount is 17. $a = 17$.

**6.** Find the percent $p$, base $b$, and amount $a$.

   **(a)** What is 23% of 524?

   **(b)** What percent of 4 is 37?

---

**7.** Find 260% of 40.

The percent $p = 260$. The number that is the base usually appears after the word "of." The base $b = 40$. The amount is unknown. We use the variable $a$. Thus,

$$\frac{a}{b} = \frac{p}{100} \text{ becomes } \frac{a}{40} = \frac{260}{100}.$$

If we simplify the fraction on the right-hand side, we have the following.

$$\frac{a}{40} = \frac{13}{5}$$
$$5a = (40)(13)$$
$$5a = 520$$
$$\frac{5a}{5} = \frac{520}{5}$$
$$a = 104$$

Thus 260% of 40 is 104.

**8.** Find 540% of 50.

| Example | Student Practice |
|---|---|
| **9.** 65% of what is 195? | **10.** 76% of what is 228? |

**9.** 65% of what is 195?

The percent $p = 65$. The base is unknown. We use the variable $b$. The amount $a$ is 195. Thus,

$$\frac{a}{b} = \frac{p}{100} \text{ becomes } \frac{195}{b} = \frac{65}{100}.$$

If we simplify the fraction on the right-hand side, we have the following.

$$\frac{195}{b} = \frac{13}{20}$$
$$(20)(195) = 13b$$
$$3900 = 13b$$
$$\frac{3900}{13} = \frac{13b}{13}$$
$$300 = b$$

Thus 65% of 300 is 195.

**10.** 76% of what is 228?

---

**11.** 19 is what percent of 95?

The percent is unknown. We use the variable $p$. The base $b = 95$. The amount $a = 19$. Thus,

$$\frac{a}{b} = \frac{p}{100} \text{ becomes } \frac{19}{95} = \frac{p}{100}.$$

Cross-multiplying, we have the following.

$$(100)(19) = 95p$$
$$1900 = 95p$$
$$\frac{1900}{95} = \frac{95p}{95}$$
$$20 = p$$

**12.** 175 is what percent of 140?

| Example | Student Practice |
|---|---|
| **13.** Sonia has $29.75 deducted from her weekly salary of $425 for a retirement plan. What percent of Sonia's salary is withheld for the retirement plan? | **14.** Vincent has $84 deducted from his bi-weekly salary of $1050 for taxes. What percent of Vincent's salary is withheld for taxes? |

We must find the percent $p$. The base $b = 425$. The amount $a = 29.75$. Thus,

$$\frac{a}{b} = \frac{p}{100} \text{ becomes } \frac{29.75}{425} = \frac{p}{100}.$$

When we cross-multiply, we obtain the following.

$$100(29.75) = 425p$$
$$2975 = 425p$$
$$\frac{2975}{425} = \frac{425p}{425}$$
$$7 = p$$

We see that 7% of Sonia's salary is deducted for the retirement plan.

## Extra Practice

**1.** Identify the percent $p$, base $b$, and amount $a$. Do not solve for the unknown. What is 88% of 198?

**2.** Find 0.02% of 950.

**3.** 1800 is 225% of what?

**4.** 96 is what percent of 240?

## Concept Check

In the following percent proportion, what can you say about the percent number if the value of the amount is larger than the base? $\dfrac{\text{amount}}{\text{base}} = \dfrac{\text{percent number}}{100}$

**Chapter 8 Decimals and Percents**
**8.9 Solving Applied Problems Involving Percents**

**Vocabulary**
commission   •   commission rate   •   interest   •   principal   •   interest rate   •   time

1. _____ is the percent used in computing the interest.

2. _____ is the amount deposited or borrowed.

3. _____ is the money earned or paid for the use of money.

4. _____ is when a person's earnings are a certain percentage of the sales he/she makes.

| **Example** | **Student Practice** |
|---|---|
| **1.** Alex is a car salesman and earns a commission rate of 9% of the price of each car he sells. If he earned $3150 commission this month, what were his total sales for the month? | **2.** Leslie is a real estate agent with a 7% commission rate. If Leslie sells a home for $274,000, what amount of commission will she earn? |

commission = commission rate × total sales

$$\$3150 = 9\% \times n$$
$$\$3150 = 0.09 \times n$$
$$\frac{\$3150}{0.09} = \frac{0.09n}{0.09}$$
$$\$35,000 = n$$

Alex's total sales were $35,000.

We will estimate to check our answer. We round 9% to 10% and then verify that 10% of his total sales ($35,000) is approximately his commission ($3150). 10% of $35,000 = \$3500$, which is close to his commission of $3150.

Vocabulary Answers: 1. interest rate  2. principal  3. interest  4. commission

| Example | Student Practice |
|---|---|
| **3.** The enrollment at Laird Elementary School was 450 students in 2011. In 2012 the enrollment decreased by 36 students. What was the percent decrease? | **4.** A suit is on sale for $75.25 off the original price of $215. By what percent is the price of the suit reduced? |

Decrease of 36 students is what percent of 450?

decrease = percent decrease × original amount

$$36 = n\% \times 450$$

$$\frac{36}{450} = n\%$$

$$0.08 = n\%$$

$$8 = n$$

The enrollment decreased by 8%.

| Example | Student Practice |
|---|---|
| **5.** Arnold earned $26,000 a year and received a 6% raise. How much is his new yearly salary? | **6.** Penny earned $37,000 a year and received a 4% raise. How much is her new yearly salary? |

First, we find the amount of the raise.

percent increase × original amount

= increase (raise)

$$6\% \times 26,000 = 0.06 \times 26,000 = \$1560$$

Now we find his new yearly salary.

original salary + raise = new salary

$$\$26,000 + 1560 = \$27,560$$

Arnold's new salary is $27,560.

Check. 6% is a little more than $\frac{1}{2}$ of 10%. Use this fact to check the answer.

| Example | Student Practice |
|---|---|
| **7.** An advertisement states that all items in a department store are reduced 30% off the original list price. What is the sale price of a flat-screen television set with a list price of $2700? | **8.** 1300 people voted in the yearly town election last year. This year the number of voters dropped by 23%. How many people voted in the town election this year? |

**7.** (continued)

First, we find the amount of the discount.

percent decrease × original amount
= decrease

$30\% \times 2700 = 0.30 \times 2700 = \$810$

Next we find the sale price.

original price − discount = sale price
$$\$2700 - \$810 = \$1890$$

The sale price is $1890.

| Example | Student Practice |
|---|---|
| **9.** Larsen borrowed $9400 from the bank at a simple interest rate of 13%. | **10.** Julia put $1500 in a savings account that pays 4% simple interest per year. |
| **(a)** Find the interest on the loan for 1 year. | **(a)** How much interest will Julia earn in 2 years? |

$P = \text{principal} = \$9400$

$R = \text{rate} = 13\%$

$T = \text{time} = 1 \text{ year}$

$I = P \times R \times T$

$I = \$9400 \times 0.13 \times 1 = \$1222$

The interest for 1 year is $1222.

**(b)** How much does Larsen pay back at the end of the year when he pays off the loan?

**(b)** How much money will be in Julia's savings account at the end of two years?

$$\$9400 + 1222 = \$10,622$$

| Example | Student Practice |
|---|---|
| **11.** Find the interest on a loan of $2500 that is borrowed at a simple interest rate of 9% for 3 months. | **12.** Find the interest on a loan of $2000 that is borrowed at a simple interest rate of 13% for 6 months. |

We must change 3 months to years since the formula requires that the time be in years: $T = 3$ months $= \dfrac{3}{12} = \dfrac{1}{4}$ year.

$I = P \times R \times T$

$I = 2500 \times 0.09 \times \dfrac{1}{4} = 225 \times \dfrac{1}{4} = 56.25$

The interest for 3 months is $56.25.

### Extra Practice

**1.** You must pay annual property tax of 1.25% of the value of your home. If your home is worth $140,000, how much must you pay in taxes each year?

**2.** Bill works as a phone solicitor and is paid an 8% commission on the amount of sales he makes. If Bill earned $280 in commissions last week, what were his total sales for the week?

**3.** Elizabeth's salary last year was $43,500. If she gets a 5.5% pay raise, what is her new salary?

**4.** Marta borrows $7500 at a simple annual interest rate of 8%. Six months later, she repays the loan. How much interest does she pay on the loan?

### Concept Check

Explain how to find simple interest on a loan of $6500 at an annual rate of 9% for a period of 4 months.

# MATH COACH

*Mastering the skills you need to do well on the test.*

Watch the MATH COACH videos in MyMathLab® or on You Tube™ while you work the problems below. These helpful hints will help you avoid making common errors on test problems.

## Subtracting Decimal Expression—Problem 9

Perform the operation indicated. $18.8 - 6.23$

> **Helpful Hint:** Add zeros at the end of each number, if necessary, so that the same number of digits appear to the right of each decimal point. Remember to line up the decimal points.

Did you change the problem to $18.80 - 6.23$ as your first step? Yes _____  No _____

If you answered No, examine your work carefully and perform this step again. This will help you avoid borrowing errors.

Did you line up the decimal points carefully when you wrote the numbers one beneath the other?
Yes _____  No _____

$$\begin{array}{r} 18.80 \\ -6.23 \\ \hline \end{array}$$

If you answered No, rewrite the problem on paper. Write out your steps and show your borrowing.

If you answered Problem 9 incorrectly, go back and rework the problem using these suggestions.

---

## Dividing a Decimal by a Decimal—Problem 13

Perform the operation indicated. $15.75 \div 3.5$

> **Helpful Hint:** First determine how many decimal places you must move the decimal point to the right in the divisor to make it a whole number. Then move the decimal point the *same number of places* to the right in the dividend. Be sure to place the decimal point in the quotient directly above the decimal point in the dividend.

Did you change 3.5 to 35, and 15.75 to 157.5 first?
Yes _____  No _____

If you answered No, stop and make this correction.

Did you remember to write the decimal point in the quotient directly above the decimal point in the dividend?
Yes _____  No _____

If you answered No, stop and make this correction.

When you performed the last step of your division, did you multiply $5 \times 35$ to obtain 175? Yes _____  No _____

If you answered No, examine your work and check each step for calculation errors.

Now go back and rework the problem using these suggestions.

## Changing Between Fractions, Decimals, and Percents—Problem 21

Fill in the blanks. Complete the table of equivalent notations.

| Fraction Form | Decimal Form | Percent Form |
|---|---|---|
| **(a)** _____ | **(b)** _____ | **(c)** 5% |

> **Helpful Hint:** For this problem, it is easier to convert the percent to a decimal first. To change a percent to a decimal, move the decimal point two places to the **left** and remove the % symbol. To change a decimal to a fraction, write the decimal part over a denominator that has a 1 and the same number of zeros as the number of decimal places.

Did you remember to write 5% as 5.0%, then move the decimal point two places to the left? Yes _____ No _____

If you answered No, go back and complete this step again. Remember that when you change a percent to a decimal, you are dividing that number by 100. That is why the decimal point moves two places to the left.

To change 0.05 to a fraction, did you determine that the denominator of the fraction must have a 1 and 2 zeros? Then did you write 5 as the numerator of the fraction? Yes _____ No _____

If you answered No, reread the Helpful Hint regarding changing a decimal to a fraction.

Did you reduce your fraction to lowest terms? Yes _____ No _____

If you answered Problem 21 incorrectly, go back and rework the problem using these suggestions.

## Solving Applied Problems Involving Percents—Problem 28

A computer is reduced 24% from the original price of $3300.
- **(a)** How much is the computer reduced in price?
- **(b)** What is the sale price?

> **Helpful Hint:** Write a simple percent statement in your own words that describes the situation in the applied problem.

For part **(a)**, were you able to write a statement such as "The discount is 24% of the original price of $3300?" Yes _____ No _____

If you answered No, reread the problem and see if you can write a similar statement.

Were you able to write the equation $n = 0.24 \times 3300$ to find the answer to part **(a)**? Yes _____ No _____

If you answered No, check your calculations to be sure that you correctly changed 24% to a decimal.

For part **(b)**, once you know how much the computer is reduced in price, did you subtract as follows?
original price − discount = sale price
Yes _____ No _____

Remember to include the dollar sign ($) as the units for both answers.

Now go back and rework the problem using these suggestions.

Name: _____     Date: _____

Instructor: _____     Section: _____

## Chapter 9 Graphing and Statistics
## 9.1 Interpreting and Constructing Graphs

### Vocabulary
pictograph • pie graphs • circle graph • sector
double-bar graph • comparison line graph

1. Two or more sets of data can be compared by using a _____.

2. Each piece of the pie or circle in a circle graph is called a _____.

3. A _____ uses a visually appropriate symbol to represent a number of items.

4. A _____ is useful for making comparisons.

| Example | Student Practice |
|---|---|
| **1.** Consider the following pictograph. | **2.** Use the pictograph in example **1** to answer the following. |

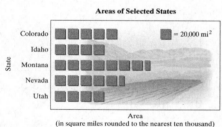

**Areas of Selected States**

Colorado ▪ = 20,000 mi²

Area
(in square miles rounded to the nearest ten thousand)

Source: World Book Encyclopedia

**(a)** How many square miles is the area of Nevada?

**(a)** How many square miles is the area of Idaho?

Since there are four symbols (▪) beside the state of Idaho on the pictograph, and each one equals 20,000 square miles, we have $4 \times 20,000 = 80,000$ square miles.

**(b)** How many more square miles is the area of Montana than the area of Nevada?

**(b)** Which of the states listed on the pictograph has the largest area?

Montana has the most symbols on the pictograph and thus has the largest area.

Vocabulary Answers: 1. comparison line graph 2. sector 3. pictograph 4. double-bar graphs

| Example | Student Practice |
|---|---|
| **3.** The total area of the five Great Lakes is about 290,000 square miles. The percentage of this total area taken up by each of the Great Lakes is shown in the pie chart. How many square miles are taken up by Lake Huron and Lake Michigan together? | **4.** Use the pie chart in example **3** to answer the following.<br><br>What percentage of the total area is not taken up by Lake Erie and Lake Ontario together? |

The percent of these two is $26 + 23 = 49$. Thus Lake Huron and Lake Michigan together take up 49% of the total area.

$$49\% \text{ of } 290,000 = (0.49)(290,000)$$
$$= 142,100 \text{ square miles}$$

| Example | Student Practice |
|---|---|
| **5.** The following double-bar graph compares the percent of all traffic fatalities for six age groups that involved drunk drivers. What percent of the traffic fatalities for males in the age range 35-44 years involved drunk drivers? | **6.** Use the double-bar graph in example **5** to answer the following.<br><br>What percent of the traffic fatalities for males in the age range 21-34 years involved drunk drivers? |

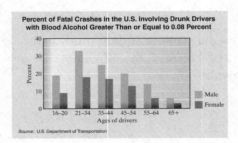

The bar rises to the white line that is halfway between 20 and 30. This represents a value halfway between 20 and 30 percent. Thus 25% of traffic fatalities for males age 35-44 involved drunk drivers.

| Example | Student Practice |
|---|---|
| **7.** Fill in the blank using the following comparison line graph. Approximately 19% of people in the _____ age group do their grocery shopping on Friday. | **8.** Use the comparison line graph in example **7** to answer the following. Fill in the blank. Approximately 13% of people in the _____ age group do their grocery on Tuesday. |

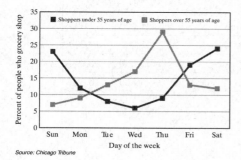

Source: Chicago Tribune

The darker line represents the under-35 age group. Since the dot corresponding to Friday on the darker line is near 20%, we know that approximately 19% of the under-35 age group shops on Friday.

| Example | Student Practice |
|---|---|

**9.** Construct a double-bar graph of the information given in the table below.

The Window Store Profits*

| Window Covering | 2011 | 2012 |
|---|---|---|
| Miniblinds | $12,000 | $15,000 |
| Vertical blinds | 11,000 | 16,000 |
| Shutters | 16,000 | 14,000 |
| Drapes | 9,000 | 7,000 |

*Profits rounded to the nearest thousand.

Since we are comparing the annual profits for the items, we place the label Profit on the vertical line. Mark intervals with equally spaced notches and label 7,000 to 17,000. We label the horizontal line with the store's products. We next label a shaded and a nonshaded bar for the years we are displaying (2011 and 2012.) Now, we draw a bar to the appropriate height for each category of window coverings.

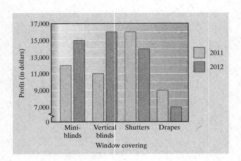

**10.** Construct a double-bar graph with the information given about a car dealership in the table below.

Car Costs

| Type | 2011 | 2012 |
|---|---|---|
| Minivans | $15,000 | $17,000 |
| Sedans | 13,000 | 19,000 |
| SUV's | 18,000 | 16,000 |
| Coupes | 11,000 | 10,000 |

## Extra Practice

1. The double-bar graph below compares the number of computer systems sold by two sales associates over four years. Over the four-year period, how many systems did Patrick sell?

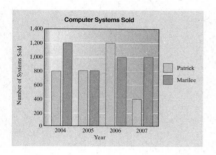

2. Use to the double-bar graph in extra practice **1** to answer the following. In 2007, how many computer systems did both sales associates sell combined?

3. Construct a comparison line graph that compares the number of birth during the first six months of the year at two local hospitals.

|                             | Jan. | Feb. | Mar. | Apr. | May | Jun. |
|-----------------------------|------|------|------|------|-----|------|
| Mercy Medical Center        | 140  | 120  | 200  | 100  | 150 | 150  |
| University Memorial Hospital| 90   | 120  | 100  | 175  | 200 | 125  |

4. Kelly earns $3000 per month as an administrative assistant. The circle graph below divides this monthly salary into basic expense categories. How many months will it take Kelly to save $900?

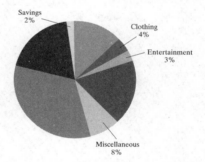

## Concept Check

The college administration is gathering data to determine the number of students who enroll in morning, afternoon, evening, and weekend courses. So far the following information has been gathered: 45% of students enroll in morning courses, while 10% enroll in weekend courses.

(a) If the enrollment in the afternoon is twice the weekend enrollment, explain how you would determine the percent of students enrolled in evening and in afternoon courses.

(b) If you create a circle graph (pie graph), describe how you would construct a pie slice that describes the percent of students enrolled in evening courses.

**Chapter 9 Graphing and Statistics**
**9.2 Mean, Median, and Mode**

**Vocabulary**

mean   •   average   •   median   •   mode   •   bimodal

1. The _____ of a set of data is the number or numbers that occur most often.

2. The _____ of a set of values is the sum of the values divided by the number of values. This is also sometimes called the average.

3. If two values occur most often, we say that the data is _____.

4. If a set of numbers is arranged in order from smallest to largest, the _____ is that value that has the same number of values above as it has below it.

| **Example** | **Student Practice** |
|---|---|
| **1.** Find the average or mean test score of a student who has test scores of 71, 83, 87, 99, 80, and 90. | **2.** Find the average or mean test score of a student who has test scores of 63, 81, 85, 97, 60, and 70. |

First find the sum of all of the student's test scores.

$$71 + 83 + 87 + 99 + 80 + 90$$

Since the total number of test scores is six, we find the mean by dividing the sum by 6.

$$\frac{71 + 83 + 87 + 99 + 80 + 90}{6}$$

$$= \frac{510}{6}$$

$$= 85$$

Thus, the mean is 85.

Vocabulary Answers: 1. mode  2. mean  3. bimodal  4. median

| Example | Student Practice |
|---|---|
| **3.** Carl and Wally each kept a log of the miles per gallon achieved by their cars for the last two months. Their results are recorded on the graph. What is the mean miles per gallon figure for the last 8 weeks for Carl? Round your answer to the nearest mile per gallon. | **4.** Use the double-line graph in example **3** to answer the following.<br><br>Find the mean miles per gallon figure for the last 8 weeks for Wally. Round your answer to the nearest mile per gallon. |

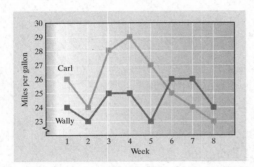

First find the sum of all of the miles per gallon figures for the last eight weeks.

$$26 + 24 + 28 + 29 + 27 + 25 + 24 + 23$$

Since Carl kept a log for eight weeks, we divide the sum by 8.

$$\frac{26 + 24 + 28 + 29 + 27 + 25 + 24 + 23}{8}$$

$$= \frac{206}{8}$$

$$\approx 26$$

The mean miles per gallon figure, rounded to the nearest whole number is 26.

| Example | Student Practice |
|---|---|
| **5.** The total minutes of daily telephone calls made by Sara and Brad during one week are indicated on the double-bar graph. Find the median value for the total minutes of Sara's daily calls. | **6.** Use the double-bar graph in example **5** to answer the following.<br><br>Find the median value for Brad's calls for Tuesday through Saturday. |

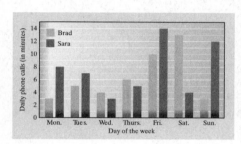

We arrange the number for Sara's calls in order from smallest to largest.

3, 4, 5          7          8, 12, 14
three numbers    ↑          three numbers
             middle
             number

There are three numbers smaller than 7 and three numbers larger than 7. Thus, 7 is the median.

**7.** Find the median of the following numbers: 13, 16, 18, 26, 31, 33, 38, and 39.

13, 16, 18      26, 31      33, 38, 39
three numbers    ↑          three numbers
             two middle
             numbers

The average (mean) of 26 and 31 is
$$\frac{26+31}{2} = \frac{57}{2} = 28.5.$$

Thus, the median value is 28.5.

**8.** Find the median of the following numbers: 86, 96, 120, 116, 146, and 128.

225

| Example | Student Practice |
|---|---|
| **9.** Find the mode of each of the following.<br><br>(a) A student's test scores of 89, 94, 96, 89, and 90.<br><br>The mode of 89, 94, 96, 89, and 90 is 89 since it occurs twice in the set of data.<br><br>(b) The ages of students in a calculus class: 33, 27, 28, 28, 21, 19, 18, 25, 26, and 33.<br><br>The data 33, 27, 28, 28, 21, 19, 18, 25, 26, and 33 are bimodal since both 28 and 33 occur twice. | **10.** Find the mode of each of the following.<br><br>(a) The number of vide rentals per day during a 1-week period: 125, 162, 135, 125, 140, 152, 129, and 140.<br><br>(b) A student's quiz score of 72, 50, 82, 95, 58, and 82. |

## Extra Practice

**1.** The annual salaries of the math department of the local community college are $64,000, $57,000, $67,000, and $63,000. Find the mean.

**2.** The ages of ten people who enter a movie theater are 35, 43, 18, 19, 28, 5, 60, 58, 39, and 33. Find the median value.

**3.** Find the mode. Over the last seven days, Holly's Plastic Surgery Center received the following numbers of inquiries: 45, 55, 65, 55, 25, 60, and 55.

**4.** A gas station listed the following gas prices over a 2-week period: $3.55, $3.28. $3.09, $3.45, $3.49, $3.70, $3.09, $3.45, $3.18, $3.60, $3.27, $3.49, $3.25, $3.57. Find the mean, median, and mode.

## Concept Check

An Internet site had the following numbers of inquiries over the last seven days: 926, 887, 778, 887, 926, 297, 801. Would you select the mean, median, or mode to determine the most realistic estimate of how many inquiries there were that particular week? Why?

Name: _____    Date: _____
Instructor: _____    Section: _____

## Chapter 9 Graphing and Statistics
## 9.3 The Rectangular Coordinate System

### Vocabulary

ordered pairs  •  rectangular coordinate system  •  axis  •  origin  •  $x$-axis
$y$-axis  •  graph  •  $x$-coordinate  •  $y$-coordinate  •  $x$-value  •  $y$-value
vertical line  •  horizontal line

1.  When all $x$-values of a set of ordered pairs are the same number, $x = a$, the coordinate points on the graph lie on a(n) _____.

2.  Each number line is called a(n) _____.

3.  The horizontal number line is often called the _____.

4.  The vertical number line is often called the _____.

| Example | Student Practice |
|---|---|
| **1.** Refer to the line graph below and list the number of applicants to medical school in the years 2003, 2005, 2007, and 2009 using ordered pairs of the form (year, number of applicants.) | **2.** Use the line graph in example **1** to answer the following. List the number of applicants to medical school in the years when the number of applicants was the lowest using ordered pairs of the form (year, number of applicants). |

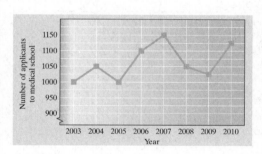

(year, number of applicants)

  ↓        ↓

(2003, 1000)

(2005, 1000)

(2007, 1150)

(2009, 1025)

Vocabulary Answers: 1. vertical line  2. axis  3. $x$-axis  4. $y$-axis

| Example | Student Practice |
|---|---|
| **3.** Plot the ordered pair $(-4,3)$. | **4.** Plot the ordered pair $(-2,6)$. |

The first number, $-4$, indicates the $x$-direction. The second number, 3, indicates the $y$-direction.

To plot $(-4,3)$ or $x = -4$, $y = 3$, we start at the origin and move 4 units in the negative $x$-direction followed by 3 units in the positive $y$-direction. We end up at the coordinate point $(-4,3)$ and place a dot there.

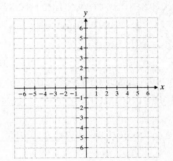

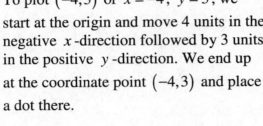

**5.** Plot the ordered pair $\left(5\frac{1}{2},1\right)$.

**6.** Plot the ordered pair $\left(0,3\frac{1}{2}\right)$.

Move $5\frac{1}{2}$ units in the positive $x$-direction. The measure of $5\frac{1}{2}$ units is located halfway between 5 and 6. Then move 1 unit up.

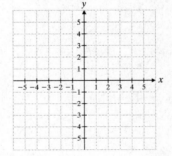

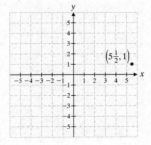

| Example | Student Practice |
|---|---|
| **7.** Give the coordinates of each point on the graph. | **8.** Give the coordinates of each point on the graph. |

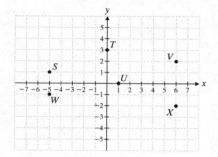

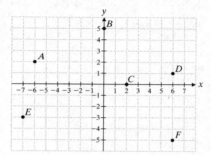

| | |
|---|---|
| **(a)** $S = (-5, 1)$ | **(a)** $A$ |
| **(b)** $T = (0, 3)$ | **(b)** $B$ |
| **(c)** $U = (1, 0)$ | **(c)** $C$ |
| **(d)** $V = (6, 2)$ | **(d)** $D$ |
| **(e)** $W = (-5, -1)$ | **(e)** $E$ |
| **(f)** $X = (6, -2)$ | **(f)** $F$ |

**9.**

**(a)** State the ordered pair that represents 3 miles west, 1 mile north.

$(-3, 1)$

**(b)** Plot the ordered pair.

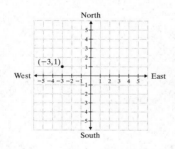

**10.**

**(a)** State the ordered pair that represents 4 miles east, 5 miles south.

**(b)** Plot the ordered pair.

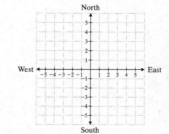

| **Example** | **Student Practice** |
|---|---|
| **11.** Plot the points corresponding to the ordered pairs and then draw a line connecting the coordinate points $(-3,4)$, $(4,4)$, $(1,4)$, $(-5,4)$. | **12.** Plot the points corresponding to the ordered pairs and then draw a line connecting the coordinate points $(4,-2)$, $(4,-5)$, $(4,0)$, $(4,6)$. |
| 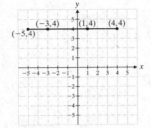 | 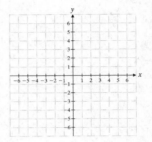 |

**Extra Practice**

**1.** Plot and label the ordered pair $(-2,-3)$ on the rectangular coordinate plane.

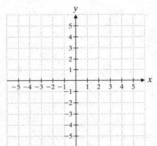

**2.** Plot and label the ordered pair $(0.5,-3)$ on the rectangular coordinate plane.

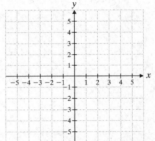

**3.** Plot the ordered pair that represents 3 miles west and 2 miles south.

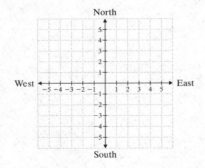

**4.** Plot the points corresponding to the set of ordered pairs and then draw a line connecting the coordinate points $(3,-2)$, $(3,0)$, $(3,-1)$, $(3,3)$.

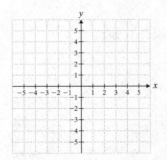

**Concept Check**

Line $M$ passes through the points $(k,b)$, $(n,b)$, and $(a,b)$. Describe line $M$.

Name: _____     Date: _____

Instructor: _____     Section: _____

## Chapter 9 Graphing and Statistics
## 9.4 Linear Equations in Two Variables

### Vocabulary

linear equation in one variable    •    linear equation in two variables
straight line    •    ordered pairs

1. The first step in the procedure to graph a line equation is to find three _____ that are solutions to the equation.

2. The graph of any linear equation in two variables is a(n) _____.

3. An equation such as $x + 2 = 7$ is called a(n) _____.

4. A(n) _____ is an equation that can be written in the form $Ax + By = C$ where $A$, $B$, and $C$ are any numbers but $A$ and $B$ are not both zero.

| Example | Student Practice |
|---|---|
| **1.** A machine at a manufacturing company can seal 50 jars per minute. The equation that represents the situation is $y = 50x$, where $y$ equals the number of jars sealed and $x$ represents the number of minutes the machine is in operation. Determine how many minutes it takes the machine to seal 200 jars. Then use your answer to write the ordered-pair solution $(x, y)$. | **2.** A machine at a manufacturing company can label 35 bottles per minute. The equation that represents the situation is $y = 35x$, where $y$ equals the number of bottles labeled and $x$ represents the number of minutes the machine is in operation. Determine how many minutes it takes the machine to label 245 bottles. Then use your answer to write the ordered pair solution $(x, y)$. |

Replace $y$ with 200 and solve for $x$.

$y = 50x$

$200 = 50x$

$\dfrac{200}{50} = x$

$4 = x$

It takes the machine 4 minutes to seal 200 jars. Since $x = 4$ when $y = 200$, the ordered pair $(x, y)$ is $(4, 200)$.

Vocabulary Answers: 1. ordered pairs  2. straight line  3. linear equation in one variable  4. linear equation in two variables

| Example | Student Practice |
|---|---|
| **3.** Fill in the ordered pair so that it is a solution to the equation $x + 2y = 10$. $$(0, \_\_)$$ The first number in the ordered pair is the $x$-value and the second number is the $y$-value. Replace $x$ with 0 and solve for $y$ in the equation $x + 2y = 10$. $$x + 2y = 10$$ $$0 + 2y = 10$$ $$2y = 10$$ $$y = 5$$ $(0, 5)$ is a solution to $x + 2y = 10$. | **4.** Fill in the ordered pair so that it is a solution to the equation $x + 4y = 32$. $$(\_\_, 2)$$ |
| **5.** Find three ordered pairs that are solutions to $y = 3x - 1$. We choose three values for $x$: 0, 1, −1, and write these values in a chart. | **6.** Find three ordered pairs that are solutions to $y = 6x - 5$. |

| (x, | y) |
|---|---|
| (0 | ) |
| (1 | ) |
| (−1 | ) |

We replace the values 0, 1, and −1 for $x$ in the given equation and solve for $y$.

$$y = 3x - 1 \qquad y = 3x - 1 \qquad y = 3x - 1$$
$$= 3(0) - 1 \qquad = 3(1) - 1 \qquad = 3(-1) - 1$$
$$= 0 - 1 \qquad = 3 - 1 \qquad = -3 - 1$$
$$y = -1 \qquad y = 2 \qquad y = -4$$

We write these values in the appropriate place in the chart.

| (x, | y) |
|---|---|
| (0, | −1) |
| (1, | 2) |
| (−1, | −4) |

232

| Example | Student Practice |
|---|---|
| **7.** Answer parts **(a)** and **(b)**. | **8.** Answer parts **(a)** and **(b)**. |

**7. (a)** Name three ordered pairs that are solutions to $y = -2x - 1$.

We choose three values for $x$: 2, 0, $-1$, and find three ordered pairs that are solutions to $y = -2x - 1$.

$y = -2x - 1$
$\quad = (-2)(2) - 1$
$y = -5$
$\quad (-5, 2)$

$y = -2x - 1$
$\quad = (-2)(0) - 1$
$y = -1$
$\quad (0, -1)$

$y = -2x - 1$
$\quad = (-2)(-1) - 1$
$y = 1$
$\quad (-1, 1)$

**8. (a)** Name three ordered pairs that are solutions to $y = -4x - 1$.

**(b)** Plot these ordered pairs on a rectangular coordinate system and draw a line through the points.

We plot these ordered pairs and draw a straight line through the points.

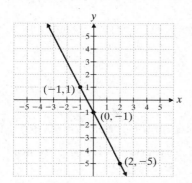

**(b)** Plot these ordered pairs on a rectangular coordinate system and draw a line through the points.

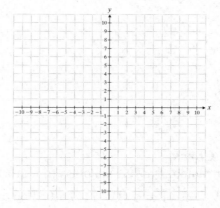

233

| Example | Student Practice |
|---|---|
| **9.** Graph $y = -1$. | **10.** Graph $x = -3$. |

A solution to $y = -1$ is any ordered pair that has $y$-coordinate $-1$. The $x$-coordinate can be any number.
The ordered pairs $(6, -1)$, $(2, -1)$, and $(-3, -1)$ are solutions to $y = -1$. We plot these ordered pairs.

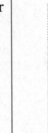

## Extra Practice

**1.** Fill in the ordered pairs so that they are solutions to the equation $x + y = 7$.

| x, | y |
|---|---|
| -2 | |
| | -4 |
| | 3 |

**2.** Find three ordered pairs that are solutions to the equation $x - y = -2$.

**3.** Find three ordered pairs that are solutions to the equation $y = 2x - 3$.

**4.** Plot three ordered-pairs solutions to the equation $y = -x - 1$ and then draw a line through the three points.

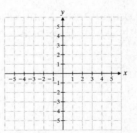

## Concept Check

Professor Sanchez asked his class to name two points that lie on the graph $y = 2x - 1$. Mark answered: $(-3, 4)$ and $(0, -1)$. Did he answer the question correctly? Why or why not?

# MATH COACH

*Mastering the skills you need to do well on the test.*

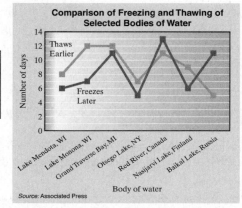

Watch the MATH COACH videos in MyMathLab® or on YouTube™ while you work the problems below. These helpful hints will help you avoid making common errors on test problems.

## Reading Circle Graphs—Problem 2

What percent of the students are between ages 18 and 24?

> **Helpful Hint:** Study the labels on the circle graph very carefully. Sometimes you need to use more than one section of the circle graph in order to answer the question.

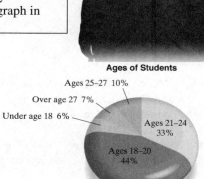

**Ages of Students**

Ages 25–27 10%
Over age 27 7%
Under age 18 6%
Ages 21–24 33%
Ages 18–20 44%

Did you get either 33% or 44% as your answer?

Yes _____ No _____

If you answered Yes, examine the circle graph carefully. Notice that the information for the age range 18-24 involves two sections of the graph. Please stop now and rework the problem.

Did you add $44\% + 33\%$ as your next step?
Yes _____ No _____
If you answered No, stop and study the graph so that you understand why the answer requires you to add these two percents.

If you answered Problem 2 incorrectly, go back and rework the problem using these suggestions.

---

## Reading and Interpreting a Comparison Line graph—Problem 7

According to this graph, which body of water froze the fewest number of days later than it did 100 years ago?

> **Helpful Hint:** Study the information on the comparison line graph carefully. Pay particular attention to the labels on each of the lines. Make sure you know which line represent the information you need to answer the question.

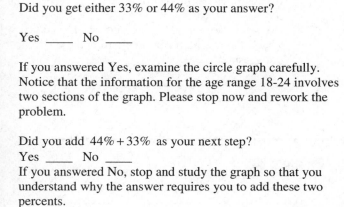

**Comparison of Freezing and Thawing of Selected Bodies of Water**

Thaws Earlier
Freezes Later
Number of days
Body of water
Lake Mendota, WI; Lake Monona, WI; Grand Traverse Bay, MI; Otsego Lake, NY; Red River, Canada; Nasjarvi Lake, Finland; Baikal Lake, Russia
*Source:* Associated Press

Did you realize that each point on the blue line refers to *the number of days later* that each body of water froze?
Yes _____ No _____

If you answered No, look at the graph carefully again and take some time to figure out what information the graph displays, then rework the problem.

Did you get Baikal Lake, Russia, as your answer?
Yes _____ No _____

If you answered Yes, study the labels and each line carefully. Notice that freezing is represented by the black line while thawing is represented by the gray line.

Now go back and rework the problem using these suggestions.

## Plotting Points Given the Coordinates—Problem 17

Plot and label the ordered pair on the rectangular coordinate system. $(3,5)$

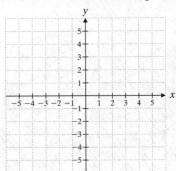

**Helpful Hint:** Remember which number in the ordered pair is $x$ and which one is $y$. To plot the points:

- Start as the origin $(0,0)$, where the $x$-axis and $y$-axis cross each other.
- Move $x$ units to the right if $x$ is positive or to the left if $x$ is negative.
- Then move $y$ units directly up if $y$ is positive or $y$ units directly down if $y$ is negative.

This is where your plotted point should be located.

To plot $(3,5)$, did you start at $(0,0)$ and move 3 units in the positive $x$-direction (right)? Yes _____ No _____

If you answered No, go back and label the $x$-coordinate. The $x$-coordinate will be the first number in your ordered pair. Stop now and rework the problem.

After moving 3 units in the positive $x$-direction, did you then move 5 units in the positive $y$-direction (up)? Yes _____ No _____

If you answered No, go back and label the $y$-coordinate. The $y$-coordinate is the second number in your ordered pair.

If you answered Problem 17 incorrectly, go back and rework the problem using these suggestions.

---

## Graphing Linear Equations in Two Variables—Problem 28

Graph $y = 3x + 1$.

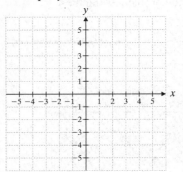

**Helpful Hint:**
- Find three ordered pairs that are solutions to the given equation.
- Be sure to replace $x$ and $y$ with the correct values.
- Verify that the three ordered pairs on the graph form a straight line.
- Double-check to make sure that you have plotted the points correctly.
- Complete Calculations very carefully.

Did you pick three ordered pairs that are solutions to $y = 3x + 1$? Yes _____ No _____

If you answered No, stop and complete this step. It helps to use a table of values to collect and organize the three ordered pairs.

When you connected the three points, did they form a straight line? Yes _____ No _____

If you answered No, check your plotted points carefully to see if they are plotted correctly. Then check your calculations for any possible errors.

Now go back and rework this problem using these suggestions.

236

**Chapter 10 Measurement and Geometric Figures**
**10.1 Using Unit Fractions with U.S. and Metric Units**

**Vocabulary**
unit fraction    •    basic unit    •    left    •    right

1.  In general, to change from a larger to a smaller metric unit, we move the decimal point to the _____.

2.  A(n) _____ is a fraction that shows the relationship between units and is equal to 1.

3.  The _____ of length in the metric system is the meter.

4.  In general, to change from a smaller to a larger metric unit, we move the decimal point to the _____.

| Example | Student Practice |
|---|---|
| **1.** Convert 35 yards to feet. | **2.** Convert 360 minutes to hours. |
| We write the relationship between feet and yards as a unit fraction. Since $3 \text{ ft} = 1 \text{ yd}$, we have the unit fraction $\dfrac{3 \text{ ft}}{1 \text{ yd}}$. | |
| $35 \text{ yd} = \underline{\ ?\ } \text{ ft}$ | |
| Now multiply by the unit fraction and then divide out the units "yd." | |
| $35 \text{ yd} \times \dfrac{3 \text{ ft}}{1 \text{ yd}}$ | |
| $= 35 \ \cancel{\text{yd}} \times \dfrac{3 \text{ ft}}{1 \ \cancel{\text{yd}}}$ | |
| $= 35 \times 3 \text{ ft} = 105 \text{ ft}$ | |

Vocabulary Answers: 1. right  2. unit fraction  3. mathematics blueprint  4. left

| Example | Student Practice |
|---|---|
| **3.** Convert 560 quarts to gallons. | **4.** Convert 128 ounces to pounds. |

**3.** Convert 560 quarts to gallons.

We write the relationship between quarts and gallons: $4 \text{ qt} = 1 \text{ gal}$. We want to end up with gallons, so we write 1 gal in the numerator of the unit fraction: $\dfrac{1 \text{ gal}}{4 \text{ qt}}$.

$560 \text{ qt} = \underline{\ ?\ } \text{ gal}$

Multiply by the appropriate unit fraction. Then divide out the units "qt."

$$560 \text{ qt} \times \frac{1 \text{ gal}}{4 \text{ qt}}$$

$$= 560 \ \cancel{\text{qt}} \times \frac{1 \text{ gal}}{4 \ \cancel{\text{qt}}}$$

$$= 560 \times \frac{1}{4} \text{ gal} = \frac{560 \text{ gal}}{4} = 140 \text{ gal}$$

---

**5.** The all-night garage charges $1.50 per hour for parking both day and night. A businessman left his car there for $2\frac{1}{4}$ days. How much was he charged?

Read the problem carefully and create a Mathematics Blueprint. We change $\frac{1}{4}$ to a decimal and change days to hours.

$$2\frac{1}{4} \text{ days} = 2.25 \ \cancel{\text{days}} \times \frac{24 \text{ hr}}{1 \ \cancel{\text{day}}} = 54 \text{ hr}$$

Now find the total charge for parking,

$$54 \ \cancel{\text{hr}} \times \frac{\$1.50}{1 \ \cancel{\text{hr}}} = \$81.$$

Check by verifying that is your answer in the desired units.

**6.** A businesswoman parked her car at a garage for $1\frac{1}{4}$ days. The garage charges $1.25 per hour. How much did she pay to park the car?

| Example | Student Practice |
|---|---|
| **7.** Answer parts **(a)** and **(b)**. | **8.** Answer parts **(a)** and **(b)**. |

**7.**

**(a)** Change 7 kilometers to meters.

To go from kilometer to meter (basic unit), we move 3 places to the right on the prefix chart, so we move the decimal point 3 places to the right.

$7 \text{ km} = 7.\underline{000} \text{ m} = 7000 \text{ m}$

**(b)** Change 30 liters to centiliters.

To go from liter (basic unit) to centiliter, we move 2 places to the right on the prefix chart. Thus we move the decimal point 2 places to the right.

$30 \text{ L} = 30.\underline{00} \text{ cL} = 3000 \text{ cL}$

**8.**

**(a)** Change 5 meters to centimeters.

**(b)** Change 40 centigrams to milligrams.

---

**9.** Answer parts **(a)** and **(b)**.

**(a)** Change 7 centigrams to grams.

To go from centigrams to grams, we move 2 places to the left on the prefix chart. Thus we move the decimal point 2 places to the left.
$7 \text{ cg} = 0.\underline{07} \text{ g} = 0.07 \text{ g}$

**(b)** Change 56 millimeters to kilometers.

To go from millimeters to kilometers, we move the decimal point 6 places to the left.

$56 \text{ mm} = 0.\underline{000056} \text{ km}$
$= 0.000056 \text{ km}$

**10.** Answer parts **(a)** and **(b)**.

**(a)** Change 5 milliliters to liters.

**(b)** Change 49 centimeters to kilometers.

| Example | Student Practice |
|---|---|
| **11.** A special cleaning fluid used to rinse test tubes in a chemistry lab costs $40.00 per liter. What is the cost per milliliter? | **12.** A purified acid costs $120 per liter. What does it cost per milliliter? |

Read the problem carefully and create a Mathematics Blueprint.

Change liters to milliliters.

$1 \text{ L} = 1000 \text{ mL}$

Replace 1 L with 1000 mL.

$$\frac{\$40}{1 \text{ L}} = \frac{\$40}{1000 \text{ mL}}$$
$$= \$0.04 \text{ per mL}$$

Check your answer. A milliliter is a very small part of a liter. Therefore it should cost much less for 1 milliliter of fluid than it does for 1 liter. $0.04 is much smaller than $40.00, so our answer seems reasonable.

**Extra Practice**

**1.** Convert. 5280 feet = __?__ miles

**2.** Convert. 1410 minutes = __?__ hours

**3.** Andrew walked 3.2 kilometers from his house to his friend's house. How many meters did he walk?

**4.** Fill in the blanks with the correct values.

0.01 L = __?__ kL = __?__ mL

**Concept Check**

Explain how you would convert 240 ounces to pounds.

**Chapter 10 Measurement and Geometric Figures**
**10.2 Converting Between the U.S. and Metric Systems**

**Vocabulary**
equivalent values   •   unit fraction   •   Celsius scale   •   Fahrenheit system

1. To convert from one unit to another we multiply by a(n) _____ that is equivalent to 1.

2. In the metric system, temperature is measured on the _____.

3. In the _____ water boils at $212°$ $(212°F)$ and freezes at $32°$ $(32°F)$.

4. To convert between U.S. and metric units, it is necessary to know _____.

| **Example** | **Student Practice** |
|---|---|
| **1.** Answer parts **(a)** through **(d)**. | **2.** Answer parts **(a)** through **(d)**. |
| **(a)** Convert 26 m to yd. | **(a)** Convert 9 ft to m. |
| $26 \text{ m} \times \dfrac{1.09 \text{ yd}}{1 \text{ m}} = 28.34 \text{ yd}$ | |
| **(b)** Convert 1.9 km to mi. | **(b)** Convert 17 qt to L. |
| $1.9 \text{ km} \times \dfrac{0.62 \text{ mi}}{1 \text{ km}} = 1.178 \text{ mi}$ | |
| **(c)** Convert 2.5 L to qt. | **(c)** Convert 22 L to gal. |
| $2.5 \text{ L} \times \dfrac{1.06 \text{ qt}}{1 \text{ L}} = 2.65 \text{ qt}$ | |
| **(d)** Convert 5.6 lb to kg. | **(d)** Convert 5 oz to g. |
| $5.6 \text{ lb} \times \dfrac{0.454 \text{ kg}}{1 \text{ lb}} = 2.5424 \text{ kg}$ | |

Vocabulary Answers: 1. unit fraction  2. Celsius scale  3. mathematics blueprint  4. equivalent values

| Example | Student Practice |
|---|---|
| **3.** Convert 235 cm to feet. Round your answer to the nearest hundredth of a foot. | **4.** Convert 194 cm to feet. Round your answer to the nearest hundredth of a foot. |

**3.** Convert 235 cm to feet. Round your answer to the nearest hundredth of a foot.

Our first unit fraction converts centimeters to inches. Our second unit fraction converts inches to feet.

$$235 \text{ cm} \times \frac{0.394 \text{ in.}}{1 \text{ cm}} \times \frac{1 \text{ ft}}{12 \text{ in.}} = \frac{92.59}{12} \text{ ft}$$
$$= 7.71583$$

Round to the nearest hundredth we have 7.72 ft.

---

**5.** Convert 100 km/hr to mi/hr.

We multiply by the unit fraction that relates mi to km.

$$\frac{100 \text{ km}}{\text{hr}} \times \frac{0.62 \text{ mi}}{1 \text{ km}} = 62 \text{ mi/hr}$$

Thus 100 km/hr is approximately equal to 62 mi/hr.

**6.** Convert 86 km/hr to mi/hr.

---

**7.** A camera film that is 35 mm wide is how many inches wide?

First convert from millimeters to centimeters by moving the decimal point in the number 35 one place to the left, 35 mm = 3.5 cm. Then convert to inches using a unit fraction.

$$3.5 \text{ cm} \times \frac{0.394 \text{ in.}}{1 \text{ cm}} = 1.379 \text{ in.}$$

**8.** The sheriff's department uses 22-mm revolvers. If such a gun fires a bullet 22 mm wide, how many inches wide is the bullet? (Round to the nearest hundredth.)

| Example | Student Practice |
|---|---|
| **9.** Convert 176°F to Celsius temperature.<br><br>Use the formula that gives us Celsius degrees.<br><br>$$C = \frac{5 \times F - 160}{9}$$<br><br>$$= \frac{5 \times 176 - 160}{9}$$<br><br>First multiply, then subtract.<br><br>$$\frac{880 - 160}{9} = \frac{720}{9} = 80$$<br><br>The temperature is 80°C. | **10.** Convert 185°F to Celsius temperature. |
| **11.** Hester is planning a visit from his home in Rhode Island to Brazil. He checked the weather report for the part of Brazil where he will visit and finds that the temperature during the day is 37°C. If the temperature in Rhode Island is currently 87°F, what is the difference between the higher and lower temperatures in degrees Fahrenheit?<br><br>Use the formula that gives us Fahrenheit degrees.<br><br>$$F = 1.8 \times C + 32$$<br><br>$$= 1.8 \times 37 + 32$$<br><br>$$= 66.6 + 32$$<br><br>$$= 98.6$$<br><br>It is 98.6°F in Brazil. Now find the difference in Fahrenheit temperatures.<br><br>$$98.6° - 87° = 11.6°F$$ | **12.** On a cold winter day in London, Drew notices that the temperature reads 2°C. She calls home to Los Angeles, California, and finds out that the temperature is 75°F. What is the difference between the higher and lower temperatures in degrees Fahrenheit? |

**Extra Practice**

1. Convert 3.2 mi to km. Round your answer to the nearest hundredth if necessary.

2. Convert 14.5 gal to L. Round your answer to the nearest hundredth if necessary.

3. Convert 45 km/hr to mi/hr. Round your answer to the nearest hundredth if necessary.

4. Convert 77°F to Celsius.

**Concept Check**

Explain how you would convert 50 km/hr to mi/hr.

Name: _____     Date: _____
Instructor: _____     Section: _____

## Chapter 10 Measurement and Geometric Figures
## 10.3 Angles

### Vocabulary
geometry • point • line • line segment • ray • angle • sides • vertex
degrees • right angle • perpendicular • straight angle • acute angle • obtuse angle
supplementary angles • complementary angles • vertical angles • adjacent angles
parallel line • transversal • alternate interior angles • corresponding angles

1. A portion of a line called a(n) _____ has a beginning and an end.

2. Two angles that are opposite of each other are called _____.

3. Two angles that have a sum of 180° are called _____.

4. Two angles that share a common side are called _____.

| Example | Student Practice |
|---|---|
| **1.** Use the figures below to answer the following. | **2.** Use the figure below to answer the following. |

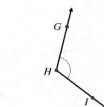
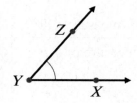

**Example**

(a) State the measure of angle *DEF* .

   Since ∠*DEF* is a straight angle, it measures 180°.

(b) State the three ways we can name the obtuse angle.

   We can name the obtuse angle as ∠*GHI*, ∠IHG, or ∠*H* . Be sure the letter representing the vertex is the middle letter.

**Student Practice**

(a) Using a single letter, name the angle in the figure.

(b) State the three ways we can name the acute angle.

Vocabulary Answers: 1. line segment  2. vertical angles  3. supplementary angles  4. adjacent angles

| Example | Student Practice |
|---|---|
| **3.** The measure of $\angle J$ is $31°$. Find the supplement of $\angle J$.<br><br>If we let $\angle S =$ the supplement of $\angle J$, then we have the following.<br><br>The supplement of $\angle J$ plus $\angle J = 180°$.<br><br>Translate this to symbols.<br><br>$\angle S + \angle J = 180°$<br><br>Now replace $\angle J$ with $31°$ and finally, solve for $\angle S$. | **4.** The measure of $\angle A$ is $24°$. Find the complement of $\angle A$. |

$$
\begin{array}{rrrr}
\angle S & + \ \angle J & = & 180° \\
\angle S & + \ 31° & = & 180° \\
+ & - \ 31° & & -31° \\
\hline
& \angle S & = & 149° \\
\end{array}
$$

The supplement of $\angle J$ measure $149°$.

---

| | |
|---|---|
| **5.** If $\angle a = 80°$, find the measure of $\angle b$.<br><br>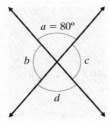<br><br>Since $\angle a$ and $\angle b$ are adjacent angles of intersecting lines, we know that they are supplementary angles. Thus, $\angle a + \angle b = 180°$. | **6.** If $\angle a = 70°$, find the measure of $\angle d$.<br><br>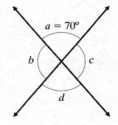 |

$$
\begin{array}{rrrr}
\angle a & + \ \angle b & = & 180° \\
80° & + \ \angle b & = & 180° \\
+ \ -80° & & & -80° \\
\hline
& \angle b & = & 100° \\
\end{array}
$$

| Example | Student Practice |
|---|---|

**7.** In the following figure $a \parallel b$, and the measure of $\angle z$ is $56°$. Find the measure of $\angle y$, $\angle x$, $\angle w$, and $\angle v$.

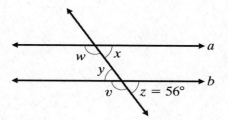

The vertical angles $z$ and $y$ have the same measure.

$\angle z = \angle y = 56°$

The alternate interior angles $y$ and $x$ have the same measure.

$\angle y = \angle x = 56°$

$\angle w$ is the supplement of $\angle x$ and $\angle x = 56°$. The sum of supplementary angles is $180°$.

$\angle w + \angle x = 180°$

Substitute $56°$ for $\angle x$ and solve for $\angle w$.

$$
\begin{array}{rcl}
\angle w \quad + \quad 56° & = & 180° \\
+ \qquad - \quad 56° & & -56° \\
\hline
\angle w & = & 124°
\end{array}
$$

The corresponding angles $w$ and $v$ have the same measure.

$\angle w = \angle v = 124°$

**8.** In the following figure $m \parallel n$, and the measure of $\angle a = 65°$. Find the measures of $\angle b$, $\angle c$, $\angle e$, and $\angle d$.

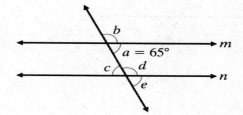

**Extra Practice**

1.  In the following figure, $p \parallel q$ and $\angle a = 68°$, find the measure of $\angle g$.

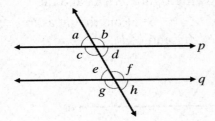

2.  Use the figure in extra practice **1** to answer the following. If $\angle a = 68°$, find the measure of $\angle a + \angle b$.

3.  In the following figure, find the measure of $\angle DBE + \angle EBC$.

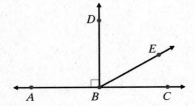

4.  Use the figure in extra practice **3** to answer the following. If $\angle ABE = 143°$ find the measure of $\angle EBC$.

**Concept Check**

In the figure shown below, explain what the relationship is between angle $a$ and angle $d$. If you know the measure of angle $a$, how can you find angle $d$?

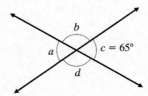

Name: _____   Date: _____
Instructor: _____   Section: _____

**Chapter 10 Measurement and Geometric Figures**
**10.4 Square Roots and the Pythagorean Theorem**

**Vocabulary**
perfect square  •  square root  •  radical sign  •  Pythagorean Theorem  •  leg
hypotenuse  •  height  •  right triangle

1.  The $\sqrt{\phantom{x}}$ symbol is called the _____.

2.  When a whole number or fraction is multiplied by itself (squared), the number obtained is called a(n) _____.

3.  The longest side which is opposite the right angle is called the _____ of the right triangle.

4.  We use the symbol $\sqrt{\phantom{x}}$ to indicate that we want to find the positive _____.

| Example | Student Practice |
|---|---|
| **1.** Determine if each number is a perfect square. | **2.** Determine if each number is a perfect square. |
| **(a)** 30 | **(a)** 23 |
| There is no whole number or fraction that when squared equals 30. Thus, 30 is not a perfect square. | |
| **(b)** 64 | **(b)** 36 |
| 64 is a perfect square because $8^2 = 8 \cdot 8 = 64$. | |
| **(c)** $\dfrac{1}{4}$ | **(c)** $\dfrac{1}{64}$ |
| $\dfrac{1}{4}$ is a perfect square because $\left(\dfrac{1}{2}\right)^2 = \dfrac{1}{2} \cdot \dfrac{1}{2} = \dfrac{1}{4}$. | |

Vocabulary Answers: 1. radical sign  2. perfect square  3. hypotenuse  4. square root

| Example | Student Practice |
|---|---|
| **3.** Write an expression for $n$ using the radical sign. Assume $n$ is a positive number.<br><br>What is the positive square root of 49?<br><br>$n = \sqrt{49}$ | **4.** Write an expression for $n$ using the radical sign. Assume $n$ is a positive number.<br><br>What is the positive square root of 144? |
| **5.** Find each square root. Then simplify if possible. $\sqrt{25} + \sqrt{36}$<br><br>$\sqrt{25} = 5$ and $\sqrt{36} = 6$. Thus,<br>$\sqrt{25} + \sqrt{36} = 5 + 6 = 11$. | **6.** Find each square root. Then simplify if possible. $\sqrt{81} - \sqrt{49}$ |
| **7.** Find the square root. $\sqrt{\dfrac{36}{49}}$<br><br>$\sqrt{\dfrac{36}{49}} = n$<br><br>What positive fraction multiplied by itself equals $\dfrac{36}{49}$?<br><br>$\sqrt{\dfrac{36}{49}} = \dfrac{6}{7}$, since $\left(\dfrac{6}{7}\right)\left(\dfrac{6}{7}\right) = \dfrac{36}{49}$. | **8.** Find the square root. $\sqrt{\dfrac{121}{169}}$ |
| **9.** Find square root of $\sqrt{27}$. Round your answer to the nearest thousandth if necessary.<br><br>Use a calculator to find $\sqrt{27}$ and round to the nearest thousandth.<br><br>$\sqrt{27} \approx 5.1961524$<br>$\qquad \approx 5.196$ | **10.** Find the square root of $\sqrt{43}$. Round your answer to the nearest thousandth if necessary. |

| Example | Student Practice |
|---|---|
| **11.** Find the area of the shape made up of a square and a right triangle, shown in the following figure. | **12.** Find the area of the shape made up of a square and a right triangle, shown in the following figure. |

10 ft

6 ft

5 ft

4 ft

We find the sum of the area of the square and the area of the triangle.

6 ft

6 ft

The area of the square is as follows.

$$A = s^2 = (6\text{ ft})^2 = 36\text{ ft}^2$$

Since all sides of a square are equal, the base of the triangle is 6 feet. We use the Pythagorean Theorem to find the height $a$ of the triangle.

$$c^2 = a^2 + b^2$$
$$10^2 = a^2 + 6^2$$
$$100 = a^2 + 36$$
$$64 = a^2$$
$$8 = a$$

$a$   10 ft

6 ft

The height of the triangle is 8 ft.

Now we find the area of the triangle. The area of the triangle is
$$A = \frac{bh}{2} = \frac{(6\text{ ft})(8\text{ ft})}{2} = 24\text{ ft}^2.$$
The sum of the two areas is
$$36\text{ ft}^2 + 24\text{ ft}^2 = 60\text{ ft}^2.$$

| Example | Student Practice |
|---|---|

**13.** A 25-foot ladder is placed against the side of a building. The top of the ladder is 22 feet from the ground. What is the distance from the base of the ladder to the building? Round to the nearest tenth if necessary.

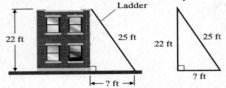

From the picture we see that we must find the length of one of the legs of a right triangle. Write the formula, replace $b$ with 22, $c$ with 25, and solve for $a$.

$$c^2 = a^2 + b^2$$
$$25^2 = a^2 + 22^2$$
$$625 = a^2 + 484$$
$$+\ -484 \qquad\qquad -\ 484$$
$$141 = a^2$$
$$\sqrt{141} = a$$
$$11.874 \approx a$$

The base of the ladder is approximately 11.9 feet from the building.

**14.** A ladder is placed against the side of a building. The top of the ladder is 17 feet from the ground. The base of the ladder is 13 feet from the building. What is the length of the ladder? Round to the nearest tenth if necessary.

## Extra Practice

**1.** Find the square root of $\sqrt{100}$.

**2.** Simplify the expression $\sqrt{625} - \sqrt{196}$.

**3.** Sheila rode her bike 5 miles south and then 2 miles east. How far is she from her starting point? Round your answer to the nearest tenth.

**4.** One leg of a right triangle measures 26 feet, and the hypotenuse measures 38 feet. Find the unknown leg. Round your answer to the nearest thousandth.

## Concept Check
State the Pythagorean Theorem and explain when you would use it.

## Chapter 10 Measurement and Geometric Figures
## 10.5 The Circle

**Vocabulary**

pi   •   center   •   circle   •   radius   •   diameter   •   circumference   •   area of a circle

1.   The distance around a circle is called its _____.

2.   The _____ of a circle is the length of a line segment from the center to a point on the circle.

3.   The _____ is the product of $\pi$ times the radius squared.

4.   A(n) _____ is a figure for which all points on the figure are at an equal distance from a given point.

| **Example** | **Student Practice** |
|---|---|
| **1.** Find the circumference of a circle if the diameter is 9.2 meters. Use $\pi \approx 3.14$. | **2.** Find the circumference of a circle if the diameter is 8.3 meters. Use $\pi \approx 3.14$. |

Since we are given the diameter, we use the formula for circumference that includes the diameter.

Write the formula.

$$C = \pi d$$

Now substitute the values given.

$$C = (3.14)(9.2 \text{ m})$$
$$C \approx 28.888 \text{ m}$$

Since $\pi$ is approximately equal to 3.14, our answer is an approximate value.

Vocabulary Answers: 1. circumference  2. radius  3. area of a circle  4. circle

| Example | Student Practice |
|---|---|
| **3.** The larger of two bicycles has a 26-inch wheel diameter, and the smaller bicycle has a 24-inch wheel diameter. | **4.** The larger of two bicycles has a 28-inch wheel diameter, and the smaller bicycle has a 22-inch wheel diameter. |

**3. (a)** If the wheels on the larger bicycle complete 12 revolutions, what distance does the larger bicycle travel?

The distance traveled by the larger bicycle during 1 revolution is equal to the measure of the circumference. Multiply the circumference by 12 to find the distance traveled in 12 revolutions.

$$C = \pi d$$
$$= (12)(\pi d)$$
$$= (12)(3.14)(26 \text{ in.})$$
$$= (12)(81.64 \text{ in.})$$
$$= 979.68 \text{ in.}$$

**(b)** How many revolutions must each wheel on the smaller bicycle complete to travel the same distance as the larger bicycle in part **(a)**?

First, find the distance traveled for 1 revolution (the circumference).

$$C = \pi d = (3.14)(24 \text{ in.}) = 75.36 \text{ in.}$$

The smaller bike travels 75.36 inches per revolution. Divide 979.68 in. by 75.36 in. to find the number of revolutions the smaller bike must make to travel the same distance as the larger bike, $979.68 \div 75.36 = 13$.

The smaller bike must complete 13 revolutions.

**4. (a)** If the wheels on the larger bicycle complete 11 revolutions, what distance does the larger bicycle travel?

**(b)** How many revolutions must each wheel on the smaller bicycle complete to travel the same distance as the larger bicycle in part **(a)**?

| Example | Student Practice |
|---|---|
| **5.** Lester wants to buy a circular braided rug that is 8 feet in diameter. Find the cost of the rug at $35 per square yard. | **6.** Jen wants to buy a circular crocheted tablecloth that is 80 inches in diameter. Find the cost of the tablecloth at $6 per square foot. |

**Example (continued):**

Read the problem carefully and create a Mathematics Blueprint.

The diameter is 8 feet, so the radius is 4 feet.

$$r = \frac{8}{2} = 4 \text{ ft}$$

Now find the area.

$$A = \pi r^2 = (3.14)(4 \text{ ft})^2$$

Square the radius first and then multiply.

$$A = (3.14)(4 \text{ ft})^2 = (3.14)\left(16 \text{ ft}^2\right)$$

$$\approx 50.24 \text{ ft}^2$$

Change square feet to square yards.
Since $1 \text{ yd} = 3 \text{ ft}$, $(1 \text{ yd})^2 = (3 \text{ ft})^2$.
That is, $1 \text{ yd}^2 = 9 \text{ ft}^2$.

$$50.24 \ \cancel{\text{ft}^2} \times \frac{1 \text{ yd}^2}{9 \ \cancel{\text{ft}^2}} \approx 5.58 \text{ yd}^2$$

(rounded to the nearest hundredth)

Finally, find the cost.

$$\frac{\$35}{1 \ \cancel{\text{yd}^2}} \times 5.58 \ \cancel{\text{yd}^2} = \$195.30$$

The rug cost $195.30.

**Extra Practice**

1. Find the circumference of the circle. Use $\pi \approx 3.14$. Round your answer to the nearest tenth.

   radius = 8.5 in.

2. A wheel makes 5 revolutions. Determine how far the bicycle travels in inches. Use $\pi \approx 3.14$. Round your answer to the nearest tenth.

   The diameter of the wheel is 42 in.

3. Find the area of the circle. Use $\pi \approx 3.14$. Round your answer to the nearest tenth.

   diameter = 3.5 in.

4. A circular, decorative window in a church has a diameter of 72 inches. Due to age, the window needs to have the insulating strip that surrounds the window replaced. How many feet of insulating strip are needed? Use $\pi \approx 3.14$. Round your answer to the nearest hundredth.

**Concept Check**

Jasmine must determine the area of a semicircle with a diameter of 10 inches. Explain how you would find the area of the semicircle.

## Chapter 10 Measurement and Geometric Figures
## 10.6 Volume

### Vocabulary
volume of a cylinder    •    volume of a sphere    •    volume of a cone
volume of a pyramid    •    volume

1. The _____ is obtained by multiplying the area of the base $B$ by the height $h$ and dividing by 3.

2. The _____ is 4 times $\pi$ times the radius cubed divided by 3.

3. The _____ is $\pi$ times the radius squared times the height divided by 3.

4. The _____ is the area of its circular base, $\pi r^2$, times the height, $h$.

| Example | Student Practice |
|---|---|
| **1.** Find the volume of a tin can with radius 7.62 centimeters and height 15.24 centimeters. Round your answer to the nearest hundredth. $$V = \pi r^2 h = (3.14)(7.62 \text{ cm})^2 (15.24 \text{ cm})$$ $$V = 2778.59 \text{ cm}^3$$ The volume of the tin can is approximately 2778.59 cubic centimeters. | **2.** Find the volume of a tin can with radius 9.45 centimeters and height 13.52 centimeters. Round your answer to the nearest hundredth. |
| **3.** How much air is needed to fully inflate a soccer ball if the radius of the inner lining is 5 inches? Round your answer to the nearest hundredth. $$V = \frac{4\pi r^3}{3} = \frac{4(3.14)(5 \text{ in.})^3}{3}$$ $$V = 523.33 \text{ in.}^3$$ Approximately 523.33 cubic inches of air is needed to fully inflate the ball. | **4.** How much air is needed to fully inflate a beach ball if the radius of the inner lining is 9 inches? Round your answer to the nearest hundredth. |

Vocabulary Answers: 1. volume of a pyramid  2. volume of a sphere  3. volume of a cone
4. volume of a cylinder

| Example | Student Practice |
|---|---|

**5.** A cylindrical thermos has a layer of insulation around its sides. The radius $R$ to the outer edge of the insulation is 15 centimeters, and the radius $r$ to the inner edge is 13 centimeters. The thermos is 27 centimeters tall. Use $\pi \approx 3.14$.

**6.** A cylindrical thermos has a layer of insulation around its sides. The radius $R$ to the outer edge of the insulation is 14 centimeters, and the radius $r$ to the inner edge is 11 centimeters. The thermos is 25 centimeters tall. Use $\pi \approx 3.14$.

**(a)** What volume of coffee can the thermos hold?

**(a)** What volume of coffee can the thermos hold?

We draw a picture of the thermos.

We find the volume of the inner shaded region $(V_r)$ to determine how much coffee the thermos will hold. The radius is 13 cm and the height is 27 cm.

$$V_r = \pi r^2 h = (3.14)(13 \text{ cm})^2 (27 \text{ cm})$$

$$V_r = 14{,}327.82 \text{ cm}^3$$

**(b)** What is the volume of the insulated region?

**(b)** What is the volume of the insulated region?

We find the volume of the entire cylinder $(V_R)$ minus the volume of the inner region $(V_r)$, $V_R - V_r$. Start by finding $V_R$.

$$V_R = \pi r^2 h = (3.14)(15 \text{ cm})^2 (27 \text{ cm})$$

$$V_R = 19{,}075.5 \text{ cm}^3$$

Now subtract to find the volume of the insulated region.

$$V_R - V_r = 19{,}075.5 - 14{,}327.82$$

$$V_R - V_r = 4747.69 \text{ cubic centimeters}$$

| Example | Student Practice |
|---|---|
| **7.** Find the volume of a cone of radius 7 meters and height 9 meters. Round to the nearest tenth. | **8.** Find the volume of a cone of radius 9 meters and height 18 meters. Round to the nearest tenth. |

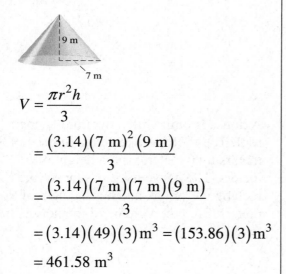

$$V = \frac{\pi r^2 h}{3}$$

$$= \frac{(3.14)(7 \text{ m})^2 (9 \text{ m})}{3}$$

$$= \frac{(3.14)(7 \text{ m})(7 \text{ m})(9 \text{ m})}{3}$$

$$= (3.14)(49)(3)\,\text{m}^3 = (153.86)(3)\,\text{m}^3$$

$$= 461.58 \text{ m}^3$$

$V \approx 461.6 \text{ m}^3$ rounded to the nearest tenth.

| **9.** Find the volume of a pyramid with height $= 6$ meters, length of base $= 7$ meters, width of base $= 5$ meters. | **10.** Find the volume of a pyramid with height $= 12$ meters, length of base $= 9$ meters, width of base $= 6$ meters. |

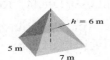

The base is a rectangle. Find its area,

Area of base $= (7 \text{ m})(5 \text{ m}) = 35 \text{ m}^2$.

Substitute the area of the base, $35 \text{ m}^2$, and the height of 6 m in the formula for the volume of the pyramid.

$$V = \frac{Bh}{3} = \frac{(35 \text{ m}^2)(6 \text{ m})}{3} = (35)(2) \text{ m}^3$$

$$= 70 \text{ m}^3$$

**Extra Practice**

1. Find the volume of a pyramid with a height of 3.5 ft and a square base of 1.5 ft on a side. Round your answer to the nearest tenth.

2. Determine the amount of air required to fill a rubber ball with a radius of 7.5 centimeters. Use $\pi \approx 3.14$. Round your answer to the nearest tenth.

3. A box is 12 inches long, 8 inches wide and 3 inches tall. It is topped by a cylinder with a diameter of 6 inches and a height of 15 inches. Find the combined volume of the box and cylinder. Use $\pi \approx 3.14$.

4. A house is built with a pyramid-shaped roof. The roof has a rectangular base of 8 meters by 6 meters, and a height of 5 meters. The house itself is shaped like a rectangular prism, and has a height of 9 meters. Find the volume of the house.

**Concept Check**

After you find the volume of a tin can with a radius of 6 inches and a height of 10 inches to be 1130.40 in.$^3$, rounded to the nearest hundredth, you decide that you must find the volume of a similar can that has all the same measurements, except the height is 5 inches. Explain how you could do this without using the formula.

Name: _____     Date: _____

Instructor: _____     Section: _____

## Chapter 10 Measurement and Geometric Figures
## 10.7 Similar Geometric Figures

### Vocabulary
similar triangles    •    corresponding angles    •    corresponding sides
perimeters    •    geometric figures

1. The length of _____ of similar triangles have the same ratio.

2. The _____ of similar triangles have the same ratios as the corresponding sides.

3. The _____ of similar triangles are equal.

4. Two triangles with the same shape but not necessarily the same size are called _____.

| Example | Student Practice |
|---|---|
| **1.** The two triangles below are similar. Find the length of side $n$. Round to the nearest tenth. | **2.** The two triangles are similar. Find the length of side $n$. Round to the nearest tenth. |

 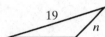

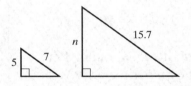

The ratio of 12 to 19 is the same as the ratio of 5 to $n$, $\dfrac{12}{19} = \dfrac{5}{n}$. Now cross-multiply and simplify.

$$12n = (19)(5)$$
$$12n = 95$$

Next, divide both sides by 12 and round to the nearest tenth.

$$\frac{12n}{12} = \frac{95}{12}$$
$$n = 7.91\overline{6}$$
$$n \approx 7.9$$

Vocabulary Answers: 1. corresponding sides  2. perimeters  3. corresponding angles  4. similar triangles

| Example | Student Practice |
|---|---|

**3.** Two triangles are similar. The smaller triangle has sides 5 yards, 7 yards, and 10 yards. The 7-yard side on the smaller triangle corresponds to a side of 21 yards on the larger triangle. What is the perimeter of the larger triangle?

We draw the two triangles.

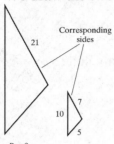

The perimeters of similar triangles have the same ratios as the corresponding sides. Therefore, we begin by finding the perimeter of the smaller triangle.

$$5 \text{ yd} + 7 \text{ yd} + 10 \text{ yd} = 22 \text{ yd}$$

We can now write equal ratios. Let $P$ = the unknown perimeter. We set up the ratios as $\dfrac{\text{smaller triangle}}{\text{larger triangle}}$. Make sure to write the terms of the second ratio in the same order.

$$\frac{22}{P} = \frac{7}{21}$$
$$7P = (21)(22)$$
$$7P = 462$$
$$\frac{7P}{7} = \frac{462}{7}$$
$$P = 66$$

The perimeter of the larger triangle is 66 yards.

**4.** Two triangles are similar. The smaller triangle has sides 7 yards, 11 yards, and 15 yards. The 15-yard side on the smaller triangle corresponds to the side of 60 yards on the larger triangle. What is the perimeter of the larger triangle?

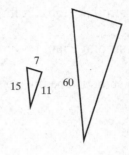

| Example | Student Practice |
|---|---|

**5.** A flagpole casts a shadow of 36 feet. At the same time, a tree that is 3 feet tall has a shadow of 5 feet. How tall is the flagpole?

The shadows cast by the sun shining on vertical objects at the same time of day form similar triangles. We draw a picture and organize our information.

Let $n$ = the height of the flagpole. Thus we can say that $n$ is to 3 as 36 is to 5.

$$\frac{n}{3} = \frac{36}{5}$$
$$5n = (3)(36)$$
$$5n = 108$$

Now divide both sides by 5, $n = 21.6$.

**6.** A flagpole casts a shadow of 30 feet. At the same time a tree that is 7.2 feet tall has a shadow of 8 feet. How tall is the flagpole?

---

**7.** The two rectangles shown below are similar because the corresponding sides of the two rectangles have the same ratio. Find the width of the larger rectangle.

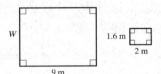

Let $W$ = the width of the larger rectangle.
$$\frac{W}{1.6} = \frac{9}{2}$$
$$2W = (1.6)(9)$$
$$2W = 14.4$$

Now divide both sides by 2, $W = 7.2$.

**8.** The two rectangles shown below are similar because the corresponding sides of the two rectangles have the same ratio. Find the width of the larger rectangle.

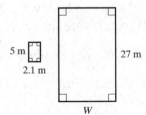

**Extra Practice**

1. For the pair of similar triangles below, find the missing side $n$. Round your answer to the nearest tenth if necessary.

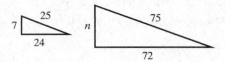

2. The pair of rectangles below are similar. Find the missing side $n$. Round to the nearest tenth if necessary.

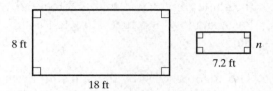

3. A tree casts a shadow of 6 feet. At the same time, a building that is 50 feet tall has a shadow that is 30 feet. How tall is the tree?

4. Keenan is rock climbing in Colorado. He is 6 ft tall and his shadow measures 9 ft long. The rock he wants to climb casts a shadow of 540 ft. How tall is the rock he wants to climb?

**Concept Check**

A tree that is 13.5 feet tall casts a shadow of 3 feet. At the same time, a building that is 45 ft tall casts a shadow of 10 feet. Using the properties of similar triangles, the height of the building is found to be 45 feet. The shadow cast by the building later in the day is $\frac{1}{2}$ of the length that it was earlier. Explain how you would find the shadow cast by the tree at the same time that afternoon.

# MATH COACH

*Mastering the skills you need to do well on the test.*

Watch the **MATH COACH** videos in MyMathLab° or on YouTube™ while you work the problems below. These helpful hints will help you avoid making common errors on test problems.

### Converting Units of Measurement—Problems 3(a) and (b)

Sharon purchased 2 kilograms of hamburger meat.
(a) How many grams did she purchase?
(b) How many pounds did she purchase?

**Helpful Hint:** Recall that 1 kilogram equals 1000 grams, and 1 kilogram is approximately 2.2 pounds.

For Part (a), did you understand that you have to move the decimal point 3 places? Yes _____ No _____

Did you remember to move the decimal point to the right? Yes _____ No _____

If you answered No to either of these questions, remember that the number of zeros in 1000 tells you how many places to move the decimal point. When converting from a larger to a smaller unit, you move the decimal point to the right.

For Part (b), did you use a unit fraction with kilograms in the denominator and pounds in the numerator? Yes _____ No _____

Did you write $2 \text{ kg} \times \left( \dfrac{2.2 \text{ lb}}{1 \text{ kg}} \right)$?

Yes _____ No _____

If you answered No to either of these questions, consider why this is the correct unit fraction and perform this step. Be careful with your calculations.

If you answered Problem 3(a) or 3(b) incorrectly, go back and rework the problems using these suggestions.

---

### Solving Applied Problems Involving Right Triangles—Problem 21

A ladder is placed against the side of a building. The top of the ladder is 14 feet from the ground. The base of the ladder is 11 feet from the building. What is the approximate length of the ladder? Round to the nearest tenth.

**Helpful Hint:** It is wise to first draw a picture and label each side of the figure. This will help you recognize that you must use the Pythagorean Theorem: $c^2 = a^2 + b^2$. Make sure you know which sides are the legs and which side is the hypotenuse: $\text{Hypotenuse}^2 = \text{leg}^2 + \left(\text{other leg}\right)^2$.

When you substituted values, did you get $c^2 = 11^2 + 14^2$? Yes _____ No _____

If you answered No, reread the problem and consider how to substitute these values correctly into the Pythagorean Theorem formula.

Did you simplify the equation to get $c^2 = 317$? Yes _____ No _____
Next, did you approximate the square root of 317 and round to the nearest tenth? Yes _____ No _____

If you answered No to either of these questions, check your calculations carefully. You can use the square root function on your calculator to approximate square roots. Remember to include the units in your final answer.

Now go back and rework this problem using these suggestions.

**Finding the Area of Circles—Problem 24** Find the area of a circle whose radius is 1.2 centimeters. Round you answer to the nearest tenth.

> **Helpful Hint:** Try to memorize all the formulas and how to use them correctly. Take the extra time to write down the formula. This will help you to avoid errors in calculations.

Did you write down the formula $A = \pi r^2$ ?
Yes _____ No _____

If you answered No, go back and complete this step.

Did you perform the calculation $3.14 \times (1.2)$ to get your final answer?
Yes _____ No _____

If you answered Yes, stop and study the formula carefully. Notice that you must square the radius, 1.2 *before you multiply*. Remember to include the units in your final answer.

If you answered Problem 24 incorrectly, go back and rework the problem using these suggestions.

---

**Finding the Volume of Pyramids—Problem 29** Find the volume of a pyramid with height 12 meters, length of base 10 meters, and width of base 7 meters.

> **Helpful Hint:** It is important that you understand how to use this formula, $V = \dfrac{Bh}{3}$ . The $B$ refers to the area of the rectangular base, and the $h$ refers to the height. You must find the area of the base first and substitute this value for $B$ in the formula.

Did you perform the calculations $10\text{ m} \times 7\text{ m}$ as your first step to find $B$ ?
Yes _____ No _____

If you answered No, stop and make this correction to find the area of the base, $B$ .

Next, did you write down $V = \dfrac{70\text{ m}^2 (12\text{ m})}{3}$ ?

Yes _____ No _____

Did you get a final unit of $\text{m}^3$ ?
Yes _____ No _____

If you answered No to either of these questions, substitute the values carefully and remember that the units must be multiplied too.

Now go back and rework the problem using these suggestions.

## 1.1

**Student Practice**

2. (a) the hundred thousands place
   (b) the thousands place
4. $5,000,000 + 300,000 + 2000 + 700$
   $+60 + 9$
6. 3 hundred-dollar bills, 8 ten-dollar bills, and 9 one-dollar bills
8. (a) eight thousand five hundred ninety-two
   (b) five million, two hundred thirty thousand, eighty-nine
10. (a) $<$
    (b) $>$
12. (a) $4 < 9$
    (b) $8 > 2$
14. 65,700
16. 5,680,000

**Extra Practice**

1. eighty thousand, fifty-nine; $80,000 + 50 + 9$
2. $>$
3. $98 < 114$
4. 2900

**Concept Check**

Answers may vary. To round 8937 to the nearest hundred, first identify the round-off place digit: 8937 . The digit to the right is less than 5. Do not change the round-off place digit. Replace all digits to the right with zeros. 8900.

## 1.2

**Student Practice**

2. (a) $6 + 2$
   (b) $x + 10$
4. The sums are shown at the top of the next column. Notice that we only need to learn 3 addition facts because the remaining are either a repeat of these or use the addition property of zero.

$6 + 0 = 6$

$5 + 1 = 6$

$4 + 2 = 6$

$3 + 3 = 6$

$2 + 4 = 6$

$1 + 5 = 6$

$0 + 6 = 6$

6. $n + 12$
8. 28
10. 1547 people
12. 21 m

**Extra Practice**

1. $a + (2 + 7); \; a + 9$
2. 1845
3. 10,951
4. 38 in.

**Concept Check**

(a) The 1 that is placed above the 9 is in the tens place. Its value is $1 \times 10$ or 10.
(b) The 1 that is placed above the 3 is in the hundreds place. Its value is $1 \times 100$ or 100.

## 1.3

**Student Practice**

2. (a) 3
   (b) 0
4. 647
6. $8 - 5$
8. 4
10. 461
12. 102 ft

**Extra Practice**

1. $400 - 301$
2. 500
3. 41,102
4. $55

**Concept Check**

Answers may vary. One possible solution follows. We can borrow only

267

from a place value that has a nonzero whole number. When we subtract 35 from 800 we cannot subtract 5 from 0, so we must change 800 to 790 and 10 ones.

## 1.4

**Student Practice**

2.

Both arrays consist of 10 items.
4. 3 and $b$ are the factors and 24 is the product.
6. $5 \cdot n = 5n$
8. 180
10. $210n$
12. 137,400
14. 340,725
16. $304

**Extra Practice**

1. Associative property of multiplication
2. $4 \cdot 2$
3. 7,528,014
4. 75 CDs

**Concept Check**

Answers may vary. One possible solution follows. Before we multiply 546 by 2000, we first separate the 2000 into its nonzero digits $(2)$ and trailing zeros $(000)$. We then multiply the nonzero digits by 546, or $2 \times 546$. To this product we then add to the right side the number of trailing zeros.

## 1.5

**Student Practice**

2. $180 \div 9$
4. $4 \div 48$
6. 4
8. (a) 1
   (b) Undefined

(c) 0
10. 315 R 30
12. 3 candy bars

**Extra Practice**

1. Undefined
2. 98 R 5
3. $52 \div x$ (Note: Choice of variable may vary.)
4. 12 plants

**Concept Check**

The following division problem has been partially solved.

$$13\overline{)2645} \quad \begin{array}{r} 2 \\ \end{array}$$

$$\underline{26}$$
$$04$$

The next step is to ask how many times the divisor $(13)$ will go into the dividend $(04)$. Because it doesn't, the next numeral in the quotient is 0.

$$13\overline{)2645} \quad \begin{array}{r} 20 \\ \end{array}$$

$$\underline{26}$$
$$04$$

Next we multiply the 0 in the quotient by the divisor $(13)$ and subtract the product from the dividend.

$$13\overline{)2645} \quad \begin{array}{r} 20 \\ \end{array}$$

$$\underline{26}$$
$$04$$
$$\underline{00}$$
$$4$$

## 1.6

**Student Practice**

2. (a) $7^8$
   (b) $9^3 a^4$
4. (a) $y \cdot y \cdot y \cdot y \cdot y \cdot y \cdot y$
   (b) $9 \cdot 9 \cdot 9 \cdot 9 \cdot 9 \cdot 9$
6. (a) 1024

(b) 1

(c) 1000

8.   1,000,000,000

10.   625

12.   (a) $12^2$

(b) $5^7$

14.   43

16.   4

**Extra Practice**

1.   $a^5$

2.   $y^6$

3.   32

4.   24

**Concept Check**

Answers may vary. One possible solution follows.

First, evaluate terms with exponents other than 1.

$50 + 3 \times 5^2 \div 25 = 50 + 3 \times 25 \div 25$

Next, evaluate multiplication operations.

$50 + 3 \times 25 \div 25 = 50 + 75 \div 25$

Next, evaluate division operations.

$50 + 75 \div 25 = 50 + 3$

Lastly, we evaluate addition operations.

$50 + 3 = 53$

## 1.7

**Student Practice**

2.   (a) $8 \cdot a + 11$

(b) $4(n + 6)$

4.   4

6.   6

8.   $32 + 4y$

10.   $9y + 41$

**Extra Practice**

1.   $8(x - 1)$

2.   5

3.   144

4.   $3x + y + 10$

**Concept Check**

Answers may vary. One possible solution is to first simplify $5(x + 1)$

using the distributive property.

$5(x + 1) = 5x + 5$

Next, evaluate the simplified expression for $x = 2$, by substituting 2 for $x$.

$5x + 5 = 5(2) + 5 = 10 + 5 = 15$

## 1.8

**Student Practice**

2.   $6b$

4.   $11x + 15y$

6.   $(x + 3y) + (6x + y) + (5x + 8y)$

$= 12x + 12y$

8.   No

10.   $n = 5$

12.   $a + 9 = 13;\ a = 4$

14.   $3x = 21;\ x = 7$

**Extra Practice**

1.   $18ab + 11$

2.   The sum of a number and eight is twelve.

3.   $a = 9$

4.   $3 + x = 12$

**Concept Check**

Answers may vary. One possible solution follows.

The first step in both processes is to combine like terms.

$3x + x + 2x = 3x + 1x + 2x$

$= (3 + 1 + 2)x = 6x$

At this point the process for (a) is complete. Because (b) has an equals sign, it can be solved by isolating the $x$. To isolate, and therefore solve for $x$, divide both sides of the equation by the coefficient of $x$.

$6x = 12$

$\dfrac{x}{6} = \dfrac{12}{6}$

$x = 2$

## 1.9

**Student Practice**

2. $22,000
4. 8250 frequent-flyer points
6. Assistant office manager

**Extra Practice**

1. Bedspread $200; Digital Camera $300; Fountain $100; Stained Glass $500
2. $1100
3. $595
4. 90 pages; 180 pages

**Concept Check**

Answers may vary. One possible solution is to list all deposits to her vacation account over the six months.

| Description of deposit | deposit amount |
| --- | --- |
| Initial balance | $200 |
| Monthly deposit $\times$ months | |
| $100 \times 6 = $600 | $600 |
| Tax return deposit $\div 2$ | |
| $900 \div 2 = $450 | $450 |
| Sum in vacation account | $1250 |

Sahara will have $1250 after six months in her vacation account. She will not have enough money to take a $1500 vacation.

## 2.1

**Student Practice**

2. ![number line from −5 to 5]

4.   (a) <
     (b) <

6.   (a) +$50
     (b) −30 ft

8. ![number line from −5 to 5 with points marked]

10.  2

12.  >

14.  −4

16.  (a) Albany
     (b) Albany, Boston, and Buffalo

**Extra Practice**

1. ![number line from −5 to 5 with points marked]

2.   21, 21

3.   42

4.   −4 has the greater opposite. The opposites are 4 and −5, and 4 is greater. (Explanations may vary.)

**Concept Check**
Answers may vary. Possible solution: The numbers are arranged left to right on the number line. They are arranged from smallest to largest.
−10, −6, −4, −1, 0

## 2.2

**Student Practice**

2.   (a)

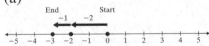

     (b) $-2+(-1)$
     (c) −3

4.   −11

6.   −7°F

8.   2

10.  18

12.  −3

14.  1

16.  $20,000

**Extra Practice**

1.   42

2.   −1

3.   −6

4.   175 feet

**Concept Check**
Answers may vary. Possible solution: Since the values given for $x$, $y$, and $z$ are all negative, their sum is negative. However, it should be obvious that the absolute value of this sum is less than 132. So when the sum of $x$, $y$, and $z$ is added to 132, we keep the sign of 132, which is positive.

## 2.3

**Student Practice**

2.   (a) −$25
     (b) −4

4.   (a) $50+(-25)=25$
     (b) $8+(-5)=3$

6.   (a) −5
     (b) −6

8.   (a) −2
     (b) −18
     (c) 10

10.  −13

12.  7

14.  3720 ft

**Extra Practice**

1.   8

2.   −3

3.   −14

4.   5:00 P.M.

**Concept Check**
Answers may vary. The problem was not completed correctly. First, subtraction is written as addition of the opposite.
$-6-(-3)+(-7)=-6+3+(-7)$

Next, the like signs are combined.
$$-6+3+(-7)=-13+3$$
Next, unlike signs are combined.
$$-13+3=-10$$

## 2.4

**Student Practice**

2.   $-32$
4.   (a) 15
     (b) $-15$
6.   240
8.   4
10.  (a) $-16$
     (b) 16
12.  (a) $-4$
     (b) 4
     (c) $-4$
14.  60
16.  (a) $-8$
     (b) $-1$

**Extra Practice**

1.   $-28$
2.   9
3.   $-1$
4.   $-8$

**Concept Check**

Answers may vary. One possible solution is it is true because an even number of negative signs occurs in the expression, the product of which must be positive.

## 2.5

**Student Practice**

2.   0
4.   8
6.   $-1$
8.   6

**Extra Practice**

1.   $-27$
2.   $-9$
3.   10
4.   $570+2(125)+4(75)$; $1120

**Concept Check**

Answers may vary. One possible

solution is to first evaluate terms with exponents.
$$3^2+5(2-4)=9+5(2-4)$$
Next, evaluate operations inside parentheses.
$$9+5(2-4)=9+5(-2)$$
Next, evaluate multiplication operations.
$$9+5(-2)=9-10$$
Lastly, combine like terms.
$$9-10=-1$$

## 2.6

**Student Practice**

2.   $-5y+5x$
4.   $4m+2n$
6.   (a) 5
     (b) $5x$
8.   $-3m-2n-5mn$
10.  $-2$
12.  $-7x+14$
14.  $s=-194$; 194 feet per second, downward

**Extra Practice**

1.   $109-3a-6ab+5b$
2.   $-96$
3.   $7a+42$
4.   $13a+1$

**Concept Check**

Answers may vary. The possible alternatives to writing $-3b+7$ are $7-3b$ or $7+(-3b)$.

## 3.1

**Student Practice**
2. 5
4. $x = -26$
6. $-9 = y$
8. (a) $\angle a = \angle b - 40°$
   (b) $100° = \angle b$
10. $x = 120°$, $\angle b = 124°$

**Extra Practice**
1. $-124$
2. $n = 38$
3. $x = 12$
4. $\angle a = 77°$

**Concept Check**
Answers may vary. It is not correct.
He should have added 9 to each side.

## 3.2

**Student Practice**
2. $-17$
4. $x = -19$
6. $y = 10$
8. $N = 7Q$
10. (a) $C = 8B$
    (b) $B = 16$
12. 17 shares

**Extra Practice**
1. $M = 7E$
2. $x = 3$
3. $-6 = x$
4. 23 feet

**Concept Check**
Answers may vary. First multiply
$3(2x)$ to get $6x$. Then combine the
like terms, $4x$ and $6x$, to get $10x$.
Finally divide each side by 10 and
simplify to get $x = -2$.

## 3.3

**Student Practice**
2. 56 yards

4. 3.75 ft
6. 54 ft$^2$
8. 81 ft$^2$
10. $b = 23$ ft
12. $W = 7$ m

**Extra Practice**
1. $P = 60$ yd; $A = 225$ yd$^2$
2. 285 in.$^2$
3. 80 ft$^3$
4. 16 ft$^3$

**Concept Check**
Answers may vary.
(a) To determine the amount of sand
    needed, find the volume.
(b) To find the volume of a box, use the
    formula $V = LWH$.
(c) Since the length is double the
    height, the length is 10 inches. The
    volume is given in cubic feet, and
    the length and height are given in
    inches. First change the units on
    these measurements so that all are
    in terms of the same unit. Then
    write an equation using the formula
    $V = LWH$, where $V$, $L$, and $H$
    are replaced with the values from
    the previous step, and solve for $W$.

## 3.4

**Student Practice**
2. $7 \cdot 7 \cdot 7 \cdot 7 \cdot 7 \cdot 7 \cdot 7 \cdot 7$, $7^8$
4. (a) $3^7$
   (b) $3^2 \cdot 2^5$
6. $168a^8b^5$
8. $-56a^3b^6$
10. (a) Trinomial
    (b) Monomial
    (c) Binomial
12. $5x^8 + 6x^7 - 24x^4$
14. $A = 5x^9 - 4x^5$

**Extra Practice**

1.  $10^{18}$
2.  $-18a^6 + 42a^5$
3.  $-4x^4 - x^3y + 20x^3 + y$
4.  $V = 60n^2 - 60n$

**Concept Check**

Answers may vary.

(a) Multiply both terms in parentheses by $2x^4$ to obtain $2x^4 \cdot x^2 + 2x^4 \cdot y$. Then add exponents on the powers of $x$ being multiplied in the first term: $2x^6 + 2x^4y$.

(b) Add the exponents on the powers of $x$ being multiplied: $-3x^6$.

(c) To complete the problem, add the simplified forms obtained in parts (a) and (b): $\left(2x^6 + 2x^4y\right) - 3x^6$. Since $2x^6$ and $-3x^6$ are like terms, complete the simplification by combining these terms: $-x^6 + 2x^4y$.

# Worksheet Answers Chapter 4

## 4.1

**Student Practice**

2.   (a) 3 and 5

    (b) 2, 3, and 5

4.   neither: 0; prime: 5, 11, 19, 29, 37, 41; composite: 9, 18, 34, 60

6.   $2 \cdot 2 \cdot 2 \cdot 2 \cdot 2 \cdot 2$ or $2^6$

8.   $2 \cdot 2 \cdot 2 \cdot 2 \cdot 5$ or $2^4 \cdot 5$

10.   $2^4 \cdot 3 \cdot 7$

12.   $2^3 \cdot 3^2$

**Extra Practice**

1.   3

2.   composite

3.   $2 \cdot 7^2$

4.   $2^5 \cdot 3^2 \cdot 5^2$

**Concept Check**

(a) Answers may vary. One possible solution is by noticing that the number ends in an even number.

(b) If any whole number ends with an even number, the number has a factor of 2.

## 4.2

**Student Practice**

2.   (a) $\dfrac{5}{6}$

    (b) $\dfrac{2}{9}$

    (c) 1

4.   (a) 1

    (b) 0

6.   $\dfrac{5}{37}$

8.   (a) Improper fraction

    (b) Improper fraction

    (c) Mixed number

10.   $7\dfrac{1}{4}$

12.   $\dfrac{49}{5}$

**Extra Practice**

1.   $\dfrac{3}{8}$

2.   $\dfrac{13}{25}$

3.   $10\dfrac{3}{8}$

4.   $\dfrac{151}{20}$

**Concept Check**

Answers may vary. One possible solution is to multiply the whole number portion by the denominator. This product is then added to the numerator.

## 4.3

**Student Practice**

2.   $\dfrac{28x}{32x}$

4.   $\dfrac{35x}{49x}$

6.   $-\dfrac{9}{4}$

8.   $\dfrac{a}{3}$

10.   (a) $\dfrac{7}{40}$

    (b) $\dfrac{59}{80}$

**Extra Practice**

1.   $\dfrac{40a}{64a}$

2.   $\dfrac{6x}{45x}$

3.   $\dfrac{5x}{8}$

4.   $\dfrac{3}{4}$

## Concept Check

Answers may vary. One quick way to determine that the fraction may be reduced is to see that both the numerator and the denominator are divisible by three. The common factor of three will cancel.

## 4.4

### Student Practice

2. (a) $5^4$

   (b) $\dfrac{1}{a^5}$

4. $\dfrac{1}{2a^3}$

6. $5^{18}$

8. (a) $5^8$

   (b) 1

   (c) $5^7 y^{21}$

10. $\dfrac{x^4}{625}$

### Extra Practice

1. $\dfrac{2b^2}{3a^2}$

2. $\dfrac{5y^8}{7z^4}$

3. $\dfrac{2^2 a^4 b^6}{c^8}$

4. $2^3 3^2 y^{12} z^4$

## Concept Check

Answers may vary. One possible approach is to simplify the rational expression in the parentheses first. The result is the quantity $2x^2$, raised to the third power. Distribution of the exponents to the quantity $2x^2$ yields $8x^6$.

## 4.5

### Student Practice

2. (a) $\dfrac{5}{7}$

(b) $\dfrac{4}{9}$

4. 19 miles per gallon

6. 46 tornados per year

8. (a) 15

   (b) 20

   (c) 16

10. (a) $7 per towel; $9 per towel

    (b) The package of 8 towels

### Extra Practice

1. $\dfrac{11}{2}$

2. $\dfrac{47}{23}$

3. $\dfrac{5}{7}$

4. pork

## Concept Check

Answers may vary. One possible solution is to write a ratio.

$$\frac{8 \text{ sales people}}{160 \text{ customers}} = \frac{1 \text{ sales person}}{20 \text{ customers}}$$

The answer is that there are 0.05 sales people per customer, or 1 sales person per 20 customers.

## 4.6

### Student Practice

2. $\dfrac{5 \text{ hours}}{150 \text{ miles}} = \dfrac{7 \text{ hours}}{210 \text{ miles}}$

4. $\dfrac{5}{17} \neq \dfrac{19}{24}$

6. This is a proportion.

8. $n = 8$

10. 56 feet

12. $3200

### Extra Practice

1. $\dfrac{4}{36} \neq \dfrac{14}{117}$

2. $x = 169$

3. 70 minutes

4. 15 books

**Concept Check**

Answers may vary. One possible method is to write a proportion. Let $x$ = amount to Justin.

$$\frac{5}{7} = \frac{x}{840}$$

$$5(840) = 7x$$

$$4200 = 7x$$

$$600 = x$$

Justin will receive $600 when Sara receives $840.

**5.1**

**Student Practice**

2.  $\dfrac{2}{5}$

4.  $\dfrac{3}{10}$

6.  $4x^7$

8.  25 ft

10. (a) $-3$

    (b) $\dfrac{1}{x}$

12. $-\dfrac{119}{195}$

14. $-\dfrac{5x^3}{18}$

16. $6\dfrac{1}{4}$ feet

**Extra Practice**

1.  $-\dfrac{3}{19}$

2.  $\dfrac{9}{35x}$

3.  $-\dfrac{3}{5}$

4.  $12x^7$

**Concept Check**

Answers may vary. One possible solution is to first write both terms with a denominator.

$$\dfrac{-16x^2}{3} \div \dfrac{8x}{1}$$

Next rewrite the expression as a multiplication problem by multiplying the first term by the reciprocal of the second term.

$$\dfrac{-16x^2}{3} \cdot \dfrac{1}{8x}$$

Multiply, and simplify.

$$-\dfrac{2x}{3}$$

**5.2**

**Student Practice**

2.  (a) $6y,\ 12y,\ 18y,\ 24y,\ 30y,$ $36y,\ 42y$

      $10y,\ 20y,\ 30y,\ 40y,\ 50y,$ $60y,\ 70y$

    (b) $30y$

4.  18

6.  2856

8.  $20x^3$

10. 2 P.M.

**Extra Practice**

1.  180

2.  42

3.  $560x^3$

4.  12:06 A.M.

**Concept Check**

Answers may vary. No, the LCM of $63x^2$ and $75x^3$ has two factors of 3 since $63x^2$ has two factors of 3 and three factors of $x$ since $75x^3$ has three factors of $x$.

**5.3**

**Student Practice**

2.  $\dfrac{13}{17}$

4.  $-\dfrac{7}{5}$

6.  (a) $\dfrac{7}{5y}$

    (b) $\dfrac{x-3}{8}$

8.  100

10. (a) 30

    (b) 5

12. $\dfrac{8}{45}$

14. $\dfrac{7x}{25}$

16. $\dfrac{8}{45}$

**Extra Practice**

1. $\dfrac{2}{15}$

2. $\dfrac{23}{50}$

3. $\dfrac{x}{6}$

4. $\dfrac{23}{40}$

**Concept Check**

(a) $20 = 2 \cdot 2 \cdot 5$

$6 = 2 \cdot 3$

$LCD = 2 \cdot 2 \cdot 3 \cdot 5 = 2^2 \cdot 3 \cdot 5 = 60$

(b) Answers may vary. One possible solution is to multiply each term by a rational number, the numerator and denominator of which are the quotient of the LCD divided by the denominator of each respective term.

## 5.4

**Student Practice**

2. $6\dfrac{6}{7}$

4. $12\dfrac{13}{20}$

6. $8\dfrac{19}{36}$

8. $1\dfrac{4}{5}$

10. $3\dfrac{1}{4}$

12. $\dfrac{156}{7}$ or $22\dfrac{2}{7}$

14. $-\dfrac{29}{24}$ or $-1\dfrac{5}{24}$

16. 8 times

**Extra Practice**

1. $14\dfrac{9}{14}$

2. $8\dfrac{13}{16}$

3. $\dfrac{481}{40}$ or $12\dfrac{1}{40}$

4. $-\dfrac{34}{3}$ or $-11\dfrac{1}{3}$

**Concept Check**

Answers may vary. One possible solution is to first express both terms as improper fractions. Multiply the numerators, multiply the denominators, then simplify.

## 5.5

**Student Practice**

2. $\dfrac{431}{875}$

4. $\dfrac{3}{128}$

6. $3x$

8. $\dfrac{45}{116}$

10. 203 inches of yarn

**Extra Practice**

1. $\dfrac{1}{125}$

2. $\dfrac{7}{100}$

3. $6x$

4. $\dfrac{1}{2}$

**Concept Check**

Answers may vary. One possible solution is to first multiply both the numerator and the denominator by 2, eliminating the denominator.

$$\dfrac{1 + 2 \times 3}{\dfrac{1}{2}} = 2(1 + 2 \times 3)$$

Next, perform the multiplication operation inside the parentheses, then add that product to 1. Multiply that sum by the 2 outside the parentheses. The expression is completely simplified.

## 5.6

**Student Practice**

2.
    (a)  $27\frac{1}{2}$ feet by $23\frac{1}{2}$ feet

    (b)  $357

4.    46 pieces of wood

**Extra Practice**

1.
    $4\frac{5}{7}$

2.
    9 pounds of meat; $16\frac{1}{5}$ pounds of

    potato salad; $13\frac{1}{2}$ pounds of fruit

3.    5 bundles of plastic tubing

4.
    $106\frac{2}{3}$ parts cement

**Concept Check**

(a) Subtract

(b) Multiply

(c) Divide

## 5.7

**Student Practice**

2.    $y = 60$

4.    $y = 6$

6.    $y = 14$

**Extra Practice**

1.    $x = 176$

2.    $x = 135$

3.    $x = -112$

4.    $x = 35$

**Concept Check**

Answers may vary. One possible
solution is that Amy intended to
eliminate the denominator. Amy failed,
however, to include the sign of the
denominator, and this error yielded an
answer with the right absolute value,
but the wrong sign.

# Worksheet Answers Chapter 6

## 6.1

### Student Practice

2. $+a^2b, +5b^2, -6a, -3b$

4. $a^2 - 8a - 1$

6. $-3x - 7y + 4z$

8. $-3x^2 + 10x - 11$

10. $-13a^2 - 53a + 24$

### Extra Practice

1. $+5a^4, -3a^3, +2a^2, -7a, +1$

2. $7y^2 - 8y - 1$

3. $4a^4 + 6a^2 - 9$

4. $5x - 6$

### Concept Check

Answers may vary. One possible error was not multiplying the negative one coefficient through the entire second expression.

## 6.2

### Student Practice

2. $-20a^2 + 40ab - 10a$

4. $-20z^7 + 40z^6 - 15z^5$

6. $8a^3 + 10a^2 + 17a + 42$

8. $x^2 + 9x + 20$

10. $y^2 + 6y + 5$

12. $x^2 - 9x + 14$

14. $7x^2 + 17x - 12$

### Extra Practice

1. $2x^3 + 5x^2 - x - 1$

2. $x^2 + 2x - 35$

3. $2x^2 + 9x - 81$

4. $-2x^3 - 5x^2 - x - 12$

### Concept Check

1. $(x+1)(x+2) = x^2 + 2x + x + 2$
$$= x^2 + 3x + 2$$

2. $(x-1)(x-2) = x^2 - 2x - x + 2$
$$= x^2 - 3x + 2$$

a. Answers may vary. One possible solution is that in number 1 all products resulting in a first order variable have a positive coefficient, whereas in number 2 the opposite is true.

b. Answers may vary. One possible solution is that all products resulting in a zero order variable are either products of two positive coefficients or two negative coefficients, both resulting in a positive coefficient.

## 6.3

### Student Practice

2. width $= W$

length $= \dfrac{1}{3}W$

4. length of the first side $= f$

length of the second side $= f + 5$

length of the third side $= 2f - 10$

6. (a) third-quarter profit $= x$

first-quarter profit $= 4x$

second-quarter profit

$= x + 25,000$

(b) $4x - 25,000$

### Extra Practice

1. price of a used vehicle $= x$

price of a new vehicle $= x + 10,100$

2. length $= L$

width $= 2L - 15$

3. length of the first side $= f$

length of the second side $= f + 4$

length of the third side $= 3f - 6$

4. (a) Sheldon's cards $= S$

Brandon's cards $= S + 350$

Trevor's cards $= S - 150$

(b) $(S+350) + (S-180) - S$

(c) $S + 170$

**Concept Check**

(a) Answers may vary. One possible solution is to let $x$ represent the height because the length and width are represented in terms of height.

(b) $x =$ height

$3x =$ width

$2x - 6 =$ length

## 6.4

**Student Practice**

2.  5

4.  5

6.  (a) $2x^2$

    (b) $y^2 z$

8.  $5(4x + 7)$

10. $6(2x + 4y - 5)$

12. $7x^2 y(2 + 5x^2 y^2)$

**Extra Practice**

1.  $x^3 y^5$

2.  $3(x - 3)$

3.  $6(5x - 2y + 3)$

4.  Not factorable; GCF is 1.

**Concept Check**

(a) Answers may vary. One possible solution is that $xy$ is not part of the GCF because $y$ is not a factor of $16x$.

(b) $12xy = 12 \cdot x \cdot y = 3 \cdot 4 \cdot x \cdot y = 2 \cdot 6 \cdot x \cdot y$

$16x = 16 \cdot x = 4 \cdot 4 \cdot x$

$GCF = 4x$

(c) $12xy + 16x = 4x \cdot 3y + 4x \cdot 4$

$\phantom{12xy + 16x} = 4x(3y + 4)$

## 7.1

**Student Practice**

2.    $-45 = y$

4.    $a = -12$

6.    $\dfrac{1}{7} = x$

8.    $x = -49$

**Extra Practice**

1.    $-8 = x$

2.    $a = 16$

3.    $x = 3$

4.    $\dfrac{5}{6} = x$

**Concept Check**

Answers may vary. Possible solution:
Equation (c) may be solved by
dividing by 7 on both sides of the
equation. This operation may be used
to isolate $x$.

## 7.2

**Student Practice**

2.    $a = -28$

4.    $x = \dfrac{10}{7}$

6.    $x = -8$

**Extra Practice**

1.    $x = \dfrac{13}{5}$

2.    $x = \dfrac{16}{5}$

3.    $x = -14$

4.    $x = 5$

**Concept Check**

Answers may vary. One possible
solution is to first combine the $x$ terms
on the left side of the equation. Then
subtract 3 from both sides in order to
combine the terms without $x$. Next
divide both sides of the equation by the
coefficient of $x$. This last step will
isolate the $x$, and solve the equation.

## 7.3

**Student Practice**

2.    $x = -6$

4.    $x = -7$

6.    $x = -8$

**Extra Practice**

1.    $x = -2$

2.    $y = -8$

3.    $x = -1$

4.    $y = -4$

**Concept Check**

(a) The first step we should perform to
solve $2(y-1)+2 = -3(y-2)$ is to
simplify the equation using the
<u>distributive</u> property.

(b) Answers may vary.
Use the distributive property.
$$2(y-1)+2 = -3(y-2)$$
$$2y-2+2 = -3y-6$$
Combine like terms.
$$2y = 3y-6$$
Subtract $3y$ from both sides to get $y$
in one term only.
$$2y-3y = 3y-3y-6$$
Simplify.
$$-y = -6$$
Divide both sides by $-1$ to isolate $y$.
$$y = 6$$

## 7.4

**Student Practice**

2.    $x = 20$

4.    $x = -\dfrac{29}{140}$

6.    $x = 5$

**Extra Practice**

1.    $x = \dfrac{1}{36}$

2.    $x = -\dfrac{3}{8}$

3.    $x = -4$

**4.** $x = \dfrac{70}{11}$

**Concept Check**

Answers may vary. One possible solution to solving the equation follows: Identify the LCD as 4, and multiply all terms.

$$-2x + \frac{3}{4} = \frac{1}{2} \rightarrow 4(-2x) + 4\left(\frac{3}{4}\right) = 4\left(\frac{1}{2}\right)$$

Multiply through to eliminate fractions.

$$4(-2x) + 4\left(\frac{3}{4}\right) = 4\left(\frac{1}{2}\right) \rightarrow -8x + 3 = 2$$

Subtract 3 from both sides of the equation.

$$-8x + 3 - 3 = 2 - 3 \rightarrow -8x = -1$$

Divide both sides by $-8$ to isolate $x$.

$$\frac{-8x}{-8} = \frac{-1}{-8} \rightarrow x = \frac{1}{8}$$

Given: $\quad S + A + M = 29$

$$A = S - 4$$

$$M = S + 2$$

Combining equations:

$$S + (S - 4) + (S + 2) = 29$$

$$3S - 2 = 29$$

$$S = \frac{31}{3} = 10\frac{1}{3}$$

$$M = S + 2 = 10\frac{1}{3} + 2 = 12\frac{1}{3}$$

Miguel completed $12\dfrac{1}{3}$ laps.

## 7.5

**Student Practice**

2. (a) 16 ft
   (b) 22 ft
4. (a) foreman's salary $= F$
       apprentice's salary $= A$
   (b) $F + (F - 6500) = 75,000$
   (c) $F = \$40,750$, $A = \$34,250$

**Extra Practice**

1. $x = 13$ m
2. $x = 4$ in.
3. Bob earns $22,000 annually.
   Carl earns $19,500 annually.
4. Sue earns $62,000 annually.
   Sam earns $31,000 annually.

**Concept Check**

(a) Answers may vary. One possible solution is that Eduardo mistakenly left out Sam's contribution to the total laps.
(b) Let $S =$ laps completed by Sam, then
    $A =$ laps completed by Alicia, and
    $M =$ laps completed by Miguel.

## 8.1

**Student Practice**

2.  (a) Seven hundred thirty-four thousandths

(b) Six and twenty-three hundredths

4.  Three hundred forty-three and $\dfrac{51}{100}$

6.  $1\dfrac{4928}{10,000}$

8.  1.023

10. >

12. 34.055

14. 151.00

**Extra Practice**

1.  Three and forty-eight hundredths

2.  19.243

3.  <

4.  3.142

**Concept Check**

Answers may vary. One possible solution follows:

There are four places after the decimal so there are four zeros in the denominator.

$8.6711 = 8\dfrac{6711}{10,000}$

## 8.2

**Student Practice**

2.  32.07

4.  27.86

6.  7.59

8.  $9.7b - 8.2a$

10. −1.32

12. $52

14. (a) 5.4 rebounds

(b) 114.6 points

**Extra Practice**

1.  28.374

2.  −6.4

3.  $2.63x + 3.4y$

4.  12.26

**Concept Check**

Answers may vary. One possible solution follows:

We replace the variable with 0.866.

$x - 3.1 = 0.866 - 3.1$

We keep the sign of the larger absolute value and subtract.

−3.100

0.866

−2.234

## 8.3

**Student Practice**

2.  94.798

4.  −39.65

6.  425,600

8.  0.45

10. −0.050

12. 4.55

14. 0.8125

16. $3.58\overline{3}$

**Extra Practice**

1.  −45.69

2.  2368.1

3.  3.25

4.  $11.2\overline{6}$

**Concept Check**

Answers may vary. One possible solution follows:

Check Marc's answer.

$0.097 \times 0.5 \overset{?}{=} 0.485$

$0.097$

$\times\ \ \ 0.5$

$0.0485$

$0.0485 \overset{?}{=} 0.485$

No, the answers are not the same.

$0.0485 \neq 0.485$

## 8.4

**Student Practice**

2. $b = 12.75$
4. $x = -5.3$
6. $x = 8.8$
8. $1.05 less
10. 7 quarters; 21 dimes

**Extra Practice**

1. $x = 6$
2. $x = 0$
3. $x = -8.125$
4. Yes, he had $13.05 left over.

**Concept Check**

Answers may vary. One possible
solution follows:
To solve the equation
$2.2 + 2.4 = 5(x-1) + 3x$:

1. Add $2.2 + 2.4$.
   $4.6 = 5(x-1) + 3x$

2. Remove parentheses.
   $4.6 = 5x - 5 + 3x$

3. Combine like terms.
   $4.6 = 8x - 5$

4. Add 5 to both sides.
   $$4.6 = 8x - 5$$
   $$\underline{+5 \qquad +5}$$
   $$9.6 = 8x$$

5. Divide both sides by 8.
   $$\frac{9.6}{8} = \frac{8x}{8}$$
   $$1.2 = x$$

## 8.5

**Student Practice**

2. (a) 7500
   (b) 750
4. (a) 320
   (b) 640
   (c) 352
6. $2
8. (a) $230
   (b) $70
   (c) $160

**Extra Practice**

1. 120
2. 240
3. 16,800
4. $14,000

**Concept Check**

Answers may vary. Two possible
solutions follow:
Find 35% of 200.

1. 35% of 200
   $= 3 \times (10\% \text{ of } 200) + 5 \times (1\% \text{ of } 200)$
   $= 3 \times 20 + 5 \times 2$
   $= 60 + 10$
   $= 70$

2. 35% of $200 = 35 \times (1\% \text{ of } 200)$
   $\qquad\qquad\qquad = 35 \times 2$
   $\qquad\qquad\qquad = 70$

## 8.6

**Student Practice**

2. 3%
4. 134%
6. 0.5%
8. (a) 0.015
   (b) 0.8%
10.

| Decimal Form | Percent Form |
|---|---|
| 0.423 | 42.3% |
| 0.644 | 64.4% |
| 0.003 | 0.3% |
| 90.1 | 9010% |

12. (a) 53.6%
    (b) $\dfrac{87}{250}$
14. $\dfrac{1}{250}$

**Extra Practice**

1. 2.45
2. 7%
3. 533.33%
4. $\dfrac{1}{800}$

## Concept Check

Answers may vary. One possible solution follows:

Change 0.43% to a decimal. Move the decimal point two places to the left.
$0.43\% = 0.0043$

Then change it to a fraction. There are 4 decimal places, so there are 4 zeros in the denominator.

$$0.43\% = 0.0043 = \frac{43}{10,000}$$

## 8.7

### Student Practice

2.   (a)  $n = 30\% \times 30;\ n = 9$

     (b)  $9 = 30\% \times n;\ n = 30$

     (c)  $9 = n\% \times 30;\ n = 30$

4.   $80 = n\% \times 50;\ n = 160$

6.   41.76

8.   81.25%

10.  16.77%

12.  $184

### Extra Practice

1.   $n = 83\% \times 155;\ n = 128.65$

2.   $240 = 80\% \times n;\ n = 300$

3.   $280 = n\% \times 70;\ n = 400$

4.   $5.21

### Concept Check

Answers may vary. One possible solution follows:

0.8% of windows are defective. There are 375 windows in a shipment. How many windows in the shipment are defective?

Let $x =$ the number of defective windows in the shipment.

$x = 0.8\%$ of 375 windows

Write the equation.

$x = 0.8\% \times 375 = 0.008 \times 375 = 3$

3 of the windows in the shipment should be defective.

## 8.8

### Student Practice

2.   (a) 20

     (b) 34

     (c)  $p$

4.   (a)  $b = 42;\ a = 8.4$

     (b)  $b$ is unknown; $a = 35$

6.   (a)  $p = 23;\ b = 524;\ a$ is unknown

     (b)  $p$ is unknown; $b = 4;\ a = 37$

8    270

10.  300

12.  125%

14.  8%

### Extra Practice

1.   $p = 88;\ b = 198;\ a$ is unknown

2.   0.19

3.   800

4.   40%

### Concept Check

Answers may vary. One possible solution follows:

In the percent proportion, what can you say about the *percent number* if the value of the *amount* is larger than the *base*?

$$\frac{\text{amount}}{\text{base}} = \frac{\text{percent number}}{100}$$

If $a$ is larger than $b$, we can write $a = b(1+n)$ where $n$ is a positive number. Substitute $b(1+n)$ for $a$.

$$\frac{b(1+n)}{b} = \frac{p}{100}$$

$$1 + n = \frac{p}{100}$$

$$100(1+n) = p$$

Since $n$ is a positive number, $(1+n)$ is larger than 1, and $p$ is larger than 100.

**Student Practice**

2.  $19,180

4.  35%

6.  $38,480

8   1001

10. (a) $120

    (b) $1620

12. $130

**Extra Practice**

1.  $1750

2.  $3500

3.  $45,892.50

4.  $300

**Concept Check**

Answers may vary. One possible solution follows:

Use the simple interest formula.

$I = P \times R \times T$

$P = \$6500, \ R = 9\%$ per year,

$T = 4$ months $= \dfrac{4}{12}$ year

$I = P \times R \times T$

$\quad = \$6500 \times 9\% \times \dfrac{4}{12}$

$\quad = 6500 \times 0.09 \times \dfrac{1}{3}$

$\quad = 195$

The interest is $195.

# Worksheet Answers Chapter 9

## 9.1

### Student Practice

2. (a) 110,000 square miles
   (b) 40,000 square miles
4. 77%
6. 33%
8. over 55
10.

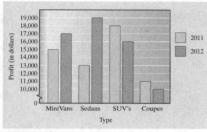

### Extra Practice

1. 3200
2. 1400
3.

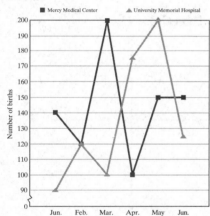

4. 15 months

### Concept Check

Answers may vary.
(a) The student enrollment is
   Morning (M) 45%
   Weekend (W) 10%
   Afternoon (A)
   $2 \times \text{weekend} = 2 \times 10\% = 20\%$
   Evening (E) The rest of the students

Total enrollment $= 100\%$

$100\% = M + W + A + E$

$100\% = 45\% + 10\% + 20\% + E$

$100\% = 75\% + E$

$\underline{-75\% \quad -75\%}$

$25\% = E$

Evening enrollment is 25% of the students.

(b) 25% is $\dfrac{1}{4}$ of the circle.

## 9.2

### Student Practice

2. 76
4. 25 miles per gallon
6. 6
8. 118
10. (a) 125 and 140
    (b) 82

### Extra Practice

1. $62,750
2. 34
3. 55
4. mean is $3.39, median is $3.45, modes are $3.09, $3.45, and $3.49

### Concept Check

Answers may vary. One possible solution follows:
First, list the number of inquiries in order.
297, 778, 801, 887, 887, 926, 926
The values fall into three "groups." One value is approximately 300, two values near 800, and four near 900. Based on these "groups," we can expect values in the mid to high 800's.
The mean may not be the best estimate because there is a very low value (297) that will affect the mean more than the other estimates, median or mode.
The median is less sensitive to high and low values. In this case, the median is 887, which agrees with the expectation

of values in the mid to high 800's.
There are two modes, 887 and 926. The
mode of 887 agrees with the median
and our expectation of mid to high
800's. However, the mode at 926 does
not. Thus, the best estimate of the
number of inquiries is the median.

(b)

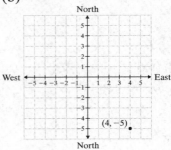

## 9.3

**Student Practice**

2. $(2003,1000)$ and $(2005,1000)$

4.

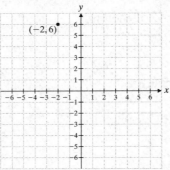

12.

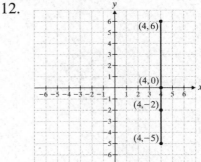

6.

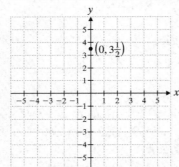

**Extra Practice**

1.

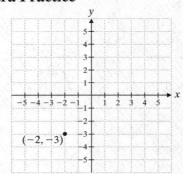

8. (a) $A = (-6,2)$

(b) $B = (0,5)$

(c) $C = (2,0)$

(d) $D = (6,1)$

(e) $E = (-7,-3)$

(f) $F = (6,-5)$

10. (a) $(4,-5)$

2.

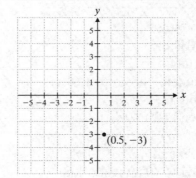

3.

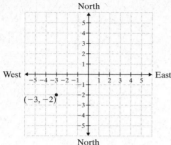

4.

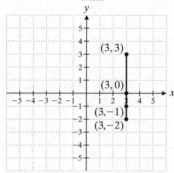

## Concept Check
Answers may vary. One possible solution follows:

Line $M$ passes through the points $(k,b)$, $(n,b)$, and $(a,b)$. Since all of the y-values are the same, the line, $M$, is parallel to the x-axis $b$ units away from it.

## 9.4

### Student Practice

2.  7 minutes, $(7,245)$

4.  $(24,2)$

6.  $(0,-5)$, $(1,1)$, $(-1,-11)$ (Answers may vary.)

8.  (a) $(0,-1)$, $(-1,3)$, $(1,-5)$
    (Answers may vary.)

(b)

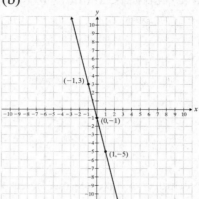

10.

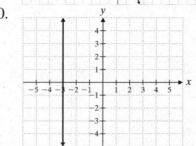

### Extra Practice

1.

| x, | y |
|----|-----|
| -2 | 9 |
| 11 | -4 |
| 4 | 3 |

2.  $(0,2)$, $(-2,0)$, $(-1,1)$ (Answers may vary.)

3.  $(0,-3)$, $(1,1)$, $(-1,-5)$ (Answers may vary.)

4.

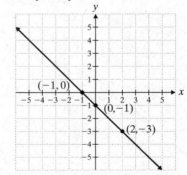

(Answers may vary.)

### Concept Check
Answers may vary. One possible solution follows:

If a point lies on a graphed line, it will be a solution to the equation of that line.

Are Mark's answers solutions to the equation $y = 2x - 1$?

$(-3, 4)$

$y \overset{?}{=} 2x - 1$

$4 \overset{?}{=} 2(-3) - 1$

$4 \overset{?}{=} -6 - 1$

$4 \neq -7$

No, the point $(-3, 4)$ does not lie on the line.

$(0, -1)$

$y \overset{?}{=} 2x - 1$

$-1 \overset{?}{=} 2(0) - 1$

$-1 \overset{?}{=} 0 - 1$

$-1 = -1$

Yes, the point $(0, -1)$ lies on the line.

So Mark's answer of the two points is not correct.

# Worksheet Answers Chapter 10

## 10.1

**Student Practice**

2. 6 hours
4. 8 pounds
6. $37.50
8. (a) 500 cm
   (b) 400 mg
10. (a) 0.005 L
    (b) 0.00049 km
12. $0.12 per mL

**Extra Practice**

1. 1 mile
2. 23.5 hours
3. 3200 meters
4. 0.01 L = 0.00001 kL = 10 mL

**Concept Check**

Answers may vary. Multiply 240 ounces by the unit fraction

$\dfrac{1\ \text{pound}}{16\ \text{ounces}}$.

## 10.2

**Student Practice**

2. (a) 2.745 m
   (b) 16.082 L
   (c) 5.808 gal
   (d) 141.75 g
4. 6.37 ft
6. 53.32 mi/hr
8. 0.8668 in.
10. 85°C
12. 39.4°F

**Extra Practice**

1. 5.15 km
2. 54.96 L
3. 27.9 mi/hr
4. 25°C

**Concept Check**

Answers may vary. Multiply $\dfrac{50\ \text{km}}{\text{hr}}$ by

the unit fraction $\dfrac{0.62\ \text{mi}}{1\ \text{km}}$.

## 10.3

**Student Practice**

2. (a) $\angle Y$
   (b) $\angle ZYX$, $\angle Y$, $\angle XYZ$
4. 66°
6. 70°
8. $\angle b = 115°$, $\angle c = 65°$,
   $\angle d = 115°$, $\angle e = 65°$

**Extra Practice**

1. 112°
2. 180°
3. 90°
4. 37°

**Concept Check**

Answers may vary. $\angle a$ and $\angle d$ are supplementary. Subtract the measure of $\angle a$ from 180° to find the measure of $\angle d$.

## 10.4

**Student Practice**

2. (a) Not a perfect square
   (b) Perfect square
   (c) Perfect square
4. $n = \sqrt{144}$
6. 2
8. $\dfrac{11}{13}$
10. 6.557
12. 22 ft$^2$
14. 21.4 ft

**Extra Practice**

1. 10
2. 11
3. 5.4 miles
4. 27.713

**Concept Check**

Answers may vary. The square of the hypotenuse of a right triangle is equal to the sum of the squares of the legs. The Pythagorean Theorem can be used to find the length of the third side of a right triangle given the lengths of two of the sides.

## 10.5

**Student Practice**
2.    $C \approx 26.062$ m
4.    (a) 759.88 in.
      (b) 14
6.    $209.33

**Extra Practice**
1.    53.4 in.
2.    659.4 in.
3.    9.6 in.$^2$
4.    18.84 ft

**Concept Check**
Answers may vary. First find the radius

of the semicircle using $r = \dfrac{d}{2}$. Then find

the area of a full circle with radius $r$
using $A = \pi r^2$. Finally, divide this area
by 2 to find the area of the semicircle.

## 10.6

**Student Practice**
2.    3791.14 cm$^3$
4.    3052.08 in.$^3$
6.    (a) 9498.5 cm$^3$
      (b) 5887.5 cm$^3$
8.    1526.0 m$^3$
10.  216 m$^3$

**Extra Practice**
1.    2.6 ft$^3$
2.    1766.3 cm$^3$
3.    711.9 in.$^3$
4.    512 m$^3$

**Concept Check**
Answers may vary. Since 5 inches is
one-half of 10 inches, divide the
volume of the can, 1130.40 in.$^3$, by 2.

## 10.7

**Student Practice**
2.    $n \approx 11.2$
4.    132 yd
6.    27 ft
8.    11.34 m

**Extra Practice**
1.    $n = 21$
2.    $n = 3.2$ ft
3.    10 ft
4.    360 ft

**Concept Check**
Answers may vary. Multiply the length
of the tree's shadow earlier in the day,

3 feet, by $\dfrac{1}{2}$ to find the length of its

shadow that afternoon.